OXFORD GEOGRAPHICAL AND
ENVIRONMENTAL STUDIES

Editors: Gordon Clark, Andrew Goudie, and Ceri Peach

AN UNCOOPERATIVE COMMODITY

Editorial Advisory Board

Professor Kay Anderson (United Kingdom)
Professor Felix Driver (United Kingdom)
Professor Rita Gardner (United Kingdom)
Professor Avijit Gupta (United Kingdom)
Professor Christian Kesteloot (Belgium)
Professor David Thomas (United Kingdom)
Professor B. L. Turner II (USA)
Professor Michael Watts (USA)
Professor James Wescoat (USA)

ALSO PUBLISHED BY
OXFORD UNIVERSITY PRESS
IN THE OXFORD GEOGRAPHICAL AND
ENVIRONMENTAL STUDIES SERIES

Island Epidemics
Andrew Cliff, Peter Haggett, and Matthew Smallman-Raynor

Pension Fund Capitalism
Gordon L. Clark

Class, Ethnicity, and Community in Southern Mexico Oaxaca's Peasantries
Colin Clarke

Cultivated Landscapes of Native North America
William E. Doolittle
Paperback

Cultivated Landscapes of Native Amazonia and the Andes
William M. Denevan

Globalization and Integrated Area Development in European Cities
Frank Moulaert
Paperback

Cultivated Landscapes of Middle America on the Eve of Conquest
Thomas M. Whitmore and B. L. Turner II

Globalization and Urban Change
Capital, Culture, and Pacific Rim Mega-Projects
Kris Olds
Paperback

Conflict, Consensus, and Rationality in Environmental Planning
An Institutional Discourse Approach
Yvonne Rydin

The Globalized City
Economic Restructuring and Social Polarization in European Cities
Frank Moulaert, Araritxa Rodríguez, and Erik Swyngedouw

An Uncooperative Commodity: Privatizing Water in England and Wales

Karen J. Bakker

UNIVERSITY PRESS

OXFORD
UNIVERSITY PRESS

Great Clarendon Street, Oxford OX2 6DP

Oxford University Press is a department of the University of Oxford.
It furthers the University's objective of excellence in research, scholarship,
and education by publishing worldwide in

Oxford New York

Auckland Bangkok Buenos Aires Cape Town Chennai
Dar es Salaam Delhi Hong Kong Istanbul Karachi Kolkata
Kuala Lumpur Madrid Melbourne Mexico City Mumbai Nairobi
São Paulo Shanghai Taipei Tokyo Toronto

Oxford is a registered trade mark of Oxford University Press
in the UK and in certain other countries

Published in the United States
by Oxford University Press Inc., New York

© Karen Bakker 2003

The moral rights of the author have been asserted
Database right Oxford University Press (maker)

First published 2003

All rights reserved. No part of this publication may be reproduced,
stored in a retrieval system, or transmitted, in any form or by any means,
without the prior permission in writing of Oxford University Press,
or as expressly permitted by law, or under terms agreed with the appropriate
reprographics rights organization. Enquiries concerning reproduction
outside the scope of the above should be sent to the Rights Department,
Oxford University Press, at the address above

You must not circulate this book in any other binding or cover
and you must impose this same condition on any acquirer

British Library Cataloguing in Publication Data
Data available

Library of Congress Cataloging in Publication Data
Data available
ISBN 0–19–925365–x

1 3 5 7 9 10 8 6 4 2

Typeset by Hope Services (Abingdon) Ltd.
Printed in Great Britain
on acid-free paper by
Biddles Ltd,
Guildford and King's Lynn.

To my family

EDITORS' PREFACE

Geography and environmental studies are two closely related and burgeoning fields of academic enquiry. Both have grown rapidly over the past few decades. At once catholic in its approach and yet strongly committed to a comprehensive understanding of the world, geography has focused upon the interaction between global and local phenomena. Environmental studies, on the other hand, have shared with the discipline of geography an engagement with different disciplines, addressing wide-ranging and significant environmental issues in the scientific community and the policy community. From the analysis of climate change and physical environmental processes to the cultural dislocations of post-modernism in human geography, these two fields of enquiry have been at the forefront of attempts to comprehend transformations taking place in the world, manifesting themselves as a variety of separate but inter-related spatial scales.

The Oxford Geographical and Environmental Studies series aims to reflect this diversity and engagement. Our goal is to publish the best original research in the two related fields, and, in doing so, to demonstrate the significance of geographical and environmental perspectives for understanding the contemporary world. As a consequence, our scope is deliberately international and ranges widely in terms of topics, approaches, and methodologies. Authors are welcome from all corners of the globe. We hope the series will help to redefine the frontiers of knowledge and build bridges within the fields of geography and environmental studies. We hope also that it will cement links with issues and approaches that have originated outside the strict confines of these disciplines. In doing so, our publications contribute to the frontiers of research and knowledge while representing the fruits of particular and diverse scholarly traditions.

Gordon L. Clark
Andrew Goudie
Ceri Peach

PREFACE

Over the past two decades a remarkable transformation has taken place in water supply management in England and Wales. Water supply has (albeit to a limited degree) been transformed from a monopoly service provided at subsidized rates by the state to citizens, to a commodity supplied on a competitive basis by private companies to customers. Efficiency is prioritized over social equity. Planning for growth is being supplanted by a focus on conservation and the management of scarcity. The mechanisms of this transformation, and its implications, are the subjects of this book.

Insofar as the claims of the environment are considered alongside those of consumers in the distributional analysis, and insofar as the articulations between environmental change and socio-economic restructuring are central to the historical analysis, the book presents a political ecological analysis, understood as a political economy of the environment. Given that the book positions itself in the political economic tradition, the relative paucity of attention paid to labour requires explanation. The neglect of labour as a category in the distributional analysis, which focuses on consumers and the environment, stems in part from the fact that in broadening questions of political economy to political ecology, analytic focus broadens beyond questions of production to reproduction in its broadest sense—conditions for liveable lives and sustainable futures, for humans and non-humans alike. My concern is with reproduction in its broadest sense—of the creation and maintenance of conditions for liveable lives—rather than production in a formal sense; in this perspective, labour loses its privileged position.

A more pragmatic justification may also be made, given that much excellent research is already available on the subject of privatization and labour relations. Saunders and Harris (1994), for example, placed individuals as voters, consumers, employees and shareholders at the heart of their analysis, and O'Connell-Davidson (1993) has focused on changing employment relations. Ward has examined the oppositional politics of water in England since 1989 (Ward 1996). Ernst's analysis examines the community and consumer sectors' various attempts to influence the structure of privatization and regulation (Ernst 1994). Strang (2001) has analysed the cultural anthropological dimensions of privatization from the perspective of both workers and consumers.

This book draws on their work, and that of a number of economic historians and regulatory economists, but also on empirical data gathered during the course of doctoral and post-doctoral research, employing archival research, questionnaires, data analysis, and in-depth interviews with water supply managers and workers, consumers, government representatives, regulators, and lobby groups (Bakker 1999*a*). Given the centralized system of water management in England and Wales, knowledge of water resources management is held by a few key individuals in government and in the water companies. Accordingly, I employed the corporate interview as a qualitative research method (McDowell 1991; Schoenberger 1991*a*; 1991*b*). The information derived from the interviews is embedded within the analysis and rarely attributed, and no interviewees are identified, in part to protect interviewee confidentiality, and in part because a significant portion of the material discussed during interviews was company confidential. Access to company confidential data was given to me in several cases, but publication of this material prohibited. The confidential data was used to triangulate claims made on the basis of data in the public domain. This technique is of course fraught with the dangers of misinterpretation and manipulation, whether inadvertent or not, of data. However, as the water industry and regulatory agencies place a large amount of information in the public domain, I felt that the analysis was not substantively compromised.

The interviews were semi-structured, of similar, but necessarily unstandardized format, with a predominance of open-ended questions. As opposed to standardized, closed-ended qualitative interviews, which allow for formal hypothesis testing, open-ended qualitative interviews are a means of hypothesis building. Statistical generalizability is, in this case, forsaken in favour of an evidentiary strategy supporting inductive, rather than deductive, reasoning. I agree with Erica Schoenberger's argument that this technique is particularly useful 'in periods of great economic and social change that pose new challenges to analytical categories and theoretical principles' (Schoenberger 1991*a*, 181). My approach is thus best characterized as hypothesis-building rather than hypothesis-testing, conceptualizing the key vectors and dimensions of transformation in this period of innovation with (and re-regulation of) normative codes and regulatory practice.

Within this framework, the concept of regulation occupies a central position. Resource regulation is here understood as a multi-scalar process of conflict mediation and arbitration between interests, articulated with practices of resource production and consumption. This process of regulation is undertaken by sanctioned actors—including government, politicians, regulators, companies, labour, and consumers—whose agendas, alliances, and relative bargaining power may change over time. Government-appointed regulators occupy a strategic role in the process of regulation, but are not the only actors.

In order to understand the evolution of the British regulatory framework, regulation must be examined at both the microeconomic and macroeconomic scale. The micro-politics and practices of 'real' regulation—the application of price caps, debates over the cost of capital—are examined together with the 'mode of regulation'—the set of institutions designed to stabilize and facilitate capital accumulation over time—in which they are embedded. This enables an analytic focus on the evolving articulation between the regulatory framework—the set of rules, norms and customs which regulate policy-making and management—water-supply management practices, and the structure of the water-supply industry. This is of interest given that, at the time of privatization, advocates had high hopes for the domestic success of the utility regulatory framework, and some made confident predictions about the export of this model abroad. With this context in mind, the first part of the book explains how this regulatory framework emerged. The second part of the book explores the evolution of the regulatory framework since privatization, and examines the implications for consumers, the environment, and the water supply industry.

ACKNOWLEDGEMENTS

Employees at numerous water companies, Water UK, the (former) Department of Environment, Department of Trade and Industry, Department for International Development, the Office of Water Services, the Environment Agency, and several UK-based NGOs, gave generously of their time and knowledge. The staff at the School of Geography and the Environment and librarians at many of Oxford's libraries, particularly the School of Geography and the Bodleian's Map Room, provided invaluable and invariably cheerful assistance. Eric Leinberger's much-appreciated redesigning of figures and tables improved the graphics immeasurably.

Financial support received during the course of this research from the British Academy, Rhodes Trust, St Catherine's College, St John's College, the School of Geography and the Environment (University of Oxford), Tarmac plc, the Economic Geography Research Group of the RGS/IBG, and the Graduate Studies Committee of the University of Oxford is gratefully acknowledged. Jesus College, Oxford provided a congenial home during the period in which the final version of the book was being written.

Erik Swyngedouw's unflagging enthusiasm sustained me in the early days of my doctoral research, as did his pithy commentary on my draft chapters towards the end. Tom Downing provided an invaluable base in the Environmental Change Unit. Without Sylvia Bowerbank's friendship and encouragement I would never have begun this research. I owe an enormous debt to her perspective on the academic project.

Much of the material presented in the book has been developed and refined through discussions at conferences and workshops, and I would like to thank all those who contributed to this process, in particular those who organized and attended presentations hosted by the SOAS Water Group (University College London, 1997), Institute for British Geographers (Annual Conference, University of Surrey, 1998), Centre for Urban Technology (University of Newcastle, 1998), American Association of Geographers (Annual Conference, 2000), Faculty of Environmental Studies ('Political Ecologies of Water and Cities' conference, York University, 2000), American Water Resources Association ('Water and Law' conference, Dundee, 2001), International History of Water Association ('The Role of Water in History and Development' conference, 2001), the journal *Studies in Political Economy* (Social and Political Ecology conference, Carlton University, 2002), Laboratoire Techniques Territoires Sociétés (CNRS—Ecole Nationale des Ponts et Chaussées, 2002), PRINWASS research team ('Meaningful Interdisciplinarity: Challenges and Opportunities for Water Research' workshop, University of Oxford, 2002), Municipal Services Project ('Services for All? Water, Electricity and other State Responsibilities' conference,

Acknowledgements

University of the Witswatersrand, 2002), and the Liu Centre for the Study of Global Issues ('Water and the Future of Life on Earth' conference, Simon Fraser University, 2002).

Invaluable insights and encouragement were received from many people, including (and in no particular order): Erik Swyngedouw, Matthew Gandy, Gordon Clark, Jody Emel, David Lloyd Owen, Neil Summerton, Ben Page, Melanie Feakins, Patrick Bond, David Kinnersley, David Johnstone, Andrew Leyshon, Esteban Castro, Simon Marvin, Simon Guy, Stephen Graham, Jean Shaoul, Luis Babiano, Consuelo Giansante, Giorgos Kallis, George Monbiot, Tony Allan, Gavin Bridge, Ailsa Allen, Roger Keil, Simon Dalby, Greg Albo, Andrew Biro, Alex Loftus, Bob MacDermid, David Hall, Emanuele Lobina, Colin Leys, Caspar Henderson, Sarah Hendry, Steven Renzetti, George Monbiot, Olivier Coutard and Bernard Barraqué. Material from Chapters 5 and 6 has previously been published as articles in *Economic Geography* and *Transactions of the Institute of British Geographers*; I wish to thank the editors and some of the anonymous reviewers for their helpful comments. The support of Oxford University Press, and particularly that of the series editor Anne Ashby, is much appreciated.

Permission to reproduce sections of the following articles is gratefully acknowledged: in Chapter 5, from Clark University, material from 'Privatising Water, Producing Scarcity: The Yorkshire Drought of 1995' published in 2000 in *Economic Geography*, 76/1: 4–27; in Chapter 6, from the Royal Geographical Society and Blackwells Publishers, material from 'Paying for Water: Water Charging and Equity in England and Wales' published in 2001 in the *Transactions of the Institute of British Geographers*, 26/2: 143–64; in Chapter 7 material from 'From Public to Private to ... mutual? Restructuring Water Supply Governance in England and Wales', K. Bakker, published in 2003 in *Geoforum*, Vol. 34, 359–74, © Elsevier 2003.

Most importantly, thanks are due to Philippe Le Billon, whose humour and appetite for adventure continue to enliven life, and without whose encouragement, insights, and patient proof-reading this book could not have been written.

KJB
Vancouver
February 2003

CONTENTS

List of Figures xiv
List of Tables xv
List of Abbreviations and Acronyms xvi

PART I Privatization and Commercialization of Water Supply

1. Introduction: From 'Retreat of the State' to 'Retreat of the Market'? 3
2. Water: An Uncooperative Commodity 18
3. Building the Networks 42
4. Commercializing Water Supply 74

PART II Re-regulating the Water Supply Industry

5. Privatizing Water, Producing Scarcity: The Yorkshire Drought of 1995 101
6. Thirsting for Equity: Consumers and the Contested Politics of Water Pricing 123
7. The Retreat of the Market? Re-regulation and Water Supply Industry Restructuring 145
8. At the Frontier of the Market? Re-regulating Water Supply 179

Bibliography 196
Index 221

LIST OF FIGURES

3.1	Regional water authorities, England and Wales (1989)	59
3.2	Statutory water companies, England and Wales (1989)	61
3.3	Public water supplied, England and Wales (1961–1999)	65
3.4	Employment levels (full-time equivalent), English and Welsh water industry (1977–1999)	66
3.5	The regulatory framework	70
5.1	The Yorkshire region	103
6.1	Water and sewerage charges, by WaSC region (1989–2000)	130
6.2	Average domestic water and sewerage services charges (1976–1999)	131
7.1	UK water sector share prices (1989–2002)	153
7.2	Schematic group plc corporate structure	156
7.3	Water service and supply companies, England and Wales (2000)	160

LIST OF TABLES

2.1	Water management: state hydraulic versus market environmentalist paradigms	28
2.2	Nature as an accumulation strategy	30
3.1	Area and resident connected population, water supply and sewerage services (1998/9)	47
3.2	Water supplies in 81 leading provincial towns and cities in Britain, 1801–1901	49
3.3	Reduction in number of water services suppliers, 1945–1966	52
3.4	Water regulation in the UK	71
6.1	Equalization payments/(levies) and equivalent income, selected companies (1978/9)	126
6.2	Water authority average domestic bills, 1977/8–1978/9	126
6.3	Measured versus unmeasured water supply, households, 1989–1998	132
6.4	Disconnections of domestic properties for non-payment of charges, 1984–1998	132
6.5	Average unmeasured water and sewerage charges, per household, (1999/00)	142
6.6	Ratio of 1999/00 to 1989/90 water and sewerage bills	143
6.7	Percentage rate of increase in average household unmeasured water and sewerage charges	143
6.8	Ratio of highest to lowest average regional unmeasured water and sewerage charges	143
6.9	Highest and lowest average regional household water/sewerage bill, 1988–2000	144
6.10	Standard deviation, average unmeasured regional household charges	144
7.1	Change in effective charging limits following the 1999 Periodic Review	152
7.2	Takeovers and diversification of the water and sewerage companies, 1989–2001	157
7.3	Consolidation of the water industry, 1989–2001	159

ABBREVIATIONS AND ACRONYMS

AMP	Asset Management Plan
BPU	Budget Payment Unit (a water supply meter fitted with a 'smart card')
CAPEX	Capital expenditure
CBA	Cost-benefit analysis
CPRE	Council for the Protection of Rural England
CV	Contingent valuation
DFID	Department for International Development
DGWS	Director-General of Water Services (head of Ofwat)
DEFRA	Department of the Environment, Food, and Rural Affairs (formerly DETR)
DETR	Department of the Environment, Transport and Regions (formerly DOE)
DoE[1]	Department of the Environment
DSM	Demand side management
DTI	Department for Trade and Industry
DWI	Drinking Water Inspectorate
EA	Environment Agency
EC	European Commission
EFL	External Financing Limit
EU	European Union
HMIP	Her Majesty's Inspectorate of Pollution
IRBM	Integrated River Basin Management
IPC	integrated pollution control
LRMC	long-run marginal cost
LTA	long-term average
NGO	non-governmental organization
NRA	National Rivers Authority
ODA	Overseas Development Aid
OFWAT	Office of Water Services
OPEX	operating expenditure
PSBR	Public Sector Borrowing Requirement
RPI	Retail Price Index
RPI − X = K	generic price-cap determination formula (X = efficiency factor)
RSPB	Royal Society for the Protection of Birds
RWA	Regional Water Authority
TWUL	Thames Water Utilities Ltd.
WaSC[2]	Water and sewerage Company
WCA	Water Companies Association (now merged with WSA to form 'Water UK')

WoC	Water supply-only company
WSA	Water Services Association (now merged with WCA to form 'Water UK')
WWF(N)	World Wide Fund for Nature
YWS	Yorkshire Water Services

[1] The Department known currently as the Department of the Environment, Food and Rural Affairs was previously called the Department of the Environment, and the Department of the Environment, Transport and the Regions. For consistency, I have referred to the department by the name in use during the time period in question.

[2] The term 'water companies' has been used throughout the book when collectively referring to WaSCs and WoCs. The acronyms are used only when the need arises to distinguish between the two.

PART I

Privatization and Commercialization of Water Supply

1

Introduction: From 'Retreat of the State' to 'Retreat of the Market'?

Control of water supply is an emotive subject, and privatization of water supply inevitably controversial. Around the world, reference is made to the British experiment with water privatization—in some cases as a salutary example, in others as a warning. Why is the British case of privatized water supply of such interest globally? In part, it has drawn attention because of its innovative model of privatization: vertically integrated, equity-financed, monopoly provision of water supply by private corporations. No other country has completely privatized its water supply and sewerage systems through the creation of publicly listed corporations of comparable size. The result is a unique industry: large, private monopolies organized at the scale of river basins, with some of the highest connection rates in the world. The dominance of English water supply companies in the rapidly growing global 'private sector participation' water supply sector has also drawn attention to their domestic market. Of equal importance is the unique regulatory framework applied to water (and other network utility industries); 'price cap' regulation, and comparative competition are regulatory techniques which are increasingly widely applied beyond Britain. Finally, the outcomes of privatization, and the criteria for evaluating the performance of the privatized companies, remain open to dispute.

Some analyses of water supply privatization situate privatization within a specific political context: the Conservative government's rewriting of Britain's socio-political landscape throughout the 1980s and early 1990s (see, for example, Buttle 1996; Hay 1996; Pirie 1988; Samson 1994). Other analyses focus on the economic dimensions of privatization, situating water within debates about the relative efficiency of modes of ownership and regulation of network utilities, and/or of welfare services (see, for example, Foster 1992; Martin and Parker 1997; Newbery 1999; Vickers and Yarrow 1988). In many accounts, emphasis is placed upon the 'retreat' of the state and the advance, in the wake of deregulation, of private companies and markets.

This book, in contrast, approaches privatization as a process of socio-economic and socio-environmental *re*-regulation. The term 're-regulation' is used in two senses. First, the empirical analysis demonstrates that privatization has not necessarily implied deregulation, but has been, rather, a process in which the state has reconfigured its role, and in some instances expanded its powers and administrative reach. Second, the 'British model' is understood to be dynamic, changing rapidly and significantly in the decade since 1989, often in ways that could not be foreseen by its original architects. An analysis of re-regulation, in the sense of innovation and institutional learning within the regulatory framework, is thus necessary for understanding the ongoing evolution of the structure of the water supply industry. Building on these understandings of re-regulation, successive chapters trace the evolving structure of the British water supply industry—culminating in the rather surprising proposal by two companies, in 2000, to return themselves to not-for-profit, public ownership—and explain the political, socio-economic, and environmental drivers of re-regulation. In so doing, the analysis emphasizes the interrelationship between socio-economic restructuring, political strategies, and environmental change, and focuses on the implications of privatization and the post-privatization regulatory framework for consumers, the environment, and water supply companies.

Our entry point into this analysis is a discussion of the links between changing patterns of environmental management and the political economic landscape of Britain over the past few decades. The privatization of the English and Welsh water supply sector must be understood both within the context of the government's privatization programme of the 1980s, and also with respect to the wide-ranging transformation of water management practices over the past few decades in Britain. Some aspects of water management remain unchanged: the use of watersheds as boundaries of service provision and regulation, the universal provision requirement, and much of the technology and techniques of water provision. Most aspects of water management have, however, undergone significant change: ownership, company structure, financing, management practices, labour relations, water pricing, and environmental and social policy objectives. The reasons for and mechanisms of this transformation, and its implications for consumers, environmental management, and the water supply industry, are a central focus of this book.

The changing water management paradigm

Over the past thirty years, water governance in England and Wales has undergone significant change. Throughout much of the twentieth century, the water supply industry in England and Wales[1] was run on a monopolistic basis, and

---

[1] Water management regimes in Northern Ireland and Scotland are administratively distinct from those in England and Wales, and water supply infrastructure was never privatized in these parts of Britain. The book uses the term 'English' when referring to the English and Welsh model of privatized water supply companies. It should be noted, however, that some authors use the adjective 'British' when referring to the regulatory framework, perhaps because the generic model of utility regulation from which water supply regulation derives is applied across Britain in other utility sectors (such as energy).

regulated as a public service, with the majority of infrastructure owned by governments (municipal and then national). Drinking water was supplied with the goal of universal provision. Water pricing was based on a concept of 'social equity': household supply was not metered, and bills were linked to property value, supported through cross-subsidies between consumers, and in some instances between regions and level of governments. Potable water was a key concern for the developers of water supply networks, who were keenly aware of the links between polluted water and the cholera and typhoid epidemics that so ravaged nineteenth-century cities. Water planners focused on developing new water sources such as reservoirs, to support growing demand stemming from economic and population growth. Environmental water quality, of particular relevance to surface water in densely populated river basins, received less sustained attention. A high level of debt, variable quality of management (particularly of sewage works), and sustained water pollution contributed to the continued decline of environmental water quality in Britain for decades.

Today, water supply is run as a business by private companies, who supply water to customers on a quasi-competitive basis. Productive efficiency has increased; labour levels have been dramatically reduced. Although monopolies remain largely intact, a degree of competition has been introduced. Management techniques have changed: the supply-led 'planning for growth' approach has been supplanted by a focus on conservation and an increasing emphasis on 'demand management'. Efficiency is prioritized over social equity. Metering of domestic consumers is gradually increasing; metered bills are to reflect the full cost of water supply, implying a reduction of subsidies insofar as is practicable. Water quality has improved dramatically: river water quality in Britain now appears to be at its highest level since the Industrial Revolution (EA 2001; DEFRA 2001*a*).

Privatization and commercialization

This changing water management paradigm must be situated within the progressive and generalized transformation in the political economy of service provision in Britain. Progressively throughout the nineteenth century, the cultural symbolism linking water to modernity and personal consumption patterns that characterize the 'modern' lifestyle, the public health implications of lack of access to water, and water's strategic economic and political-territorial importance were all justifications for collective management and regulation of water. Municipalities began increasingly to involve themselves in the business of water supply. The experience with *laissez-faire* management of water supply (and other utility network industries) in the nineteenth century and a notion of 'market failure' justified public ownership of water supply infrastructure, at a time when municipal corporations were extending their reach. Nationalization of water supply occurred in a manner analogous to other strategic sectors (such as coal or oil), and the nationalized water authorities were run in a manner similar to other network utilities (such as gas and telecommunications).

By the final quarter of the twentieth century, the experience with public management had given rise to a counter-critique of 'state failure'. The notion of an invariably inefficient public sector, together with the fiscal crises of the state unable or unwilling to finance expenditure, provided a justification of private ownership of infrastructure. The perceived need to improve productive efficiency stimulated the introduction of new management practices that prioritized efficiency gains over other goals, such as social equity, that had been the focus in earlier periods. This strategy was part of a more generalized trend: the progressive commercialization of a broad range of state functions and welfare services.

Commercialization entails a displacement of public sector by private sector management institutions (understood in the sociological sense of laws, rules, norms, and customs). This may involve the reworking of decision-making mechanisms (marketable permits being substituted for public policy allocation of water rights), and of management mechanisms; with, for example, public law being replaced by contract as a regulatory mechanism, and hierarchy by competition as an incentive mechanism. It is important to note that neither privatization nor commercialization necessarily implies liberalization (or the introduction of competition); the English and Welsh water companies, for example, remained monopoly suppliers after privatization.

Across the public sector in Britain, commercialization entailed the application of commercial practices (for example cost-benefit analysis) and commercial goals (for example profit maximization), as well as management and organizational styles drawn from the (often idealized) private sector. In some cases, as in the health service, commercialization entailed the introduction of markets and market-simulating techniques as allocation and decision-making mechanisms (Leys 2001). In other cases, commercialization was accompanied by privatization, as in the case of network utility industries, including water supply, where the state's fiscal imperatives (such as reducing public-sector borrowing) coincided neatly with political imperatives (challenging the power of the trade unions).

Commercialization of water supply companies, for example, began well before privatization. Legislation in the early 1980s initiated and formalized the transformation of the water industry in Britain 'from a public service to a business organization' (Penning-Rowsell and Parker 1983: 170). By the late 1980s, the nationalized industries were best characterized as 'publicly regulated, private monopolies operating on modified market principles' (Hay 1996: 53). Privatization consolidated this transformation—with political, as well as economic dimensions.

Commercialization and subsequent privatization 'thrust [water companies] into a more commercially orientated world, wherein the organization was under pressure first to show, and then to continually expand, a return on capital employed' (O'Connell-Davidson 1993: 191). This reliance on the profit motive as the primary incentive stands in distinct contrast to the pre-privatization era. In the 1970s, 'profits' were never mentioned in reference to water production; the term 'surplus', where one existed, was considered more appropriate for an industry supplying such a basic resource (Curwen 1994). Water was considered to be a service, supplied at subsidized rates to citizens. It was also a strategic resource, and as with other such resources (for example coal), security of supply and planning for growth were the primary goals of managers. Water managers enjoyed the 'quiet life' of the monopolist in a nationalized industry; throughout much of the twentieth century, water was an 'invisible' resource in England, and a 'flush and forget' attitude characterized the public's attitude (Kinnersley 1988). Post-privatization, the water industry became increasingly high-profile, and water regulation a subject of heated debate. Managers may have remained in charge of monopolies, but the 'quiet life' had forsaken them.

The definition of the term 'privatization' underlying this analysis bears scrutiny. The word 'privatization' is often used as a blanket term to refer to corporate control by private, for-profit companies (although it generally excludes 'private' management by communities). Formally, the term privatization refers to the shift in control from the public to the private sector, through a transfer of ownership or management responsibility for water supply infrastructure. More careful analyses distinguish between full privatization (divestiture—the sale of assets to the private sector), and what are termed 'private sector partnerships' (PSPs) or 'public-private partnerships'—varieties of contractual arrangements whereby private companies manage infrastructure on behalf of its public (often municipal) owners. The 'British model' falls in the former category: its public water supply authorities were privatized through divestiture (or asset sale).

November 1989: from public to private

In December 1989, the English and Welsh public water and sewerage authorities were privatized by flotation on the London Stock Exchange. In privatizing the water supply industry, the Conservative government had a decade of experience on which to draw. Beginning in 1979, a significant proportion of the nationalized industries producing basic goods upon which the daily routine of industrialized societies depends—including gas, electricity, and telecommunications—had been sold to the private sector (Marsh 1991). By 1994, over 40 corporations had been privatized through public flotation; trade sales and management buyouts were other methods of transferring public assets to the private sector (Curwen 1994).

During this period, more than 950,000 employees were transferred from the public to the private sector (Curwen 1994); many more had been laid off in the run-up to privatization. Whereas in 1979 the nationalized industries represented 11% of Britain's GDP, by 1994, they represented just over 2% (Curwen 1994). Employment in public corporations fell from just over 2 million—or just over 5% of the working population—in 1978/9, to just over 1 million in 1986/7 (Foster 1992: 106), and to just over 600,000 (full-time equivalent) in 1991/2 (Curwen 1994). Concurrently, the proportion of the population holding shares increased, while the proportion belonging to trade unions decreased. By late 1988, the number of shareholders in Britain exceeded the number of trade union members (Pirie 1988)—a historic transition heralded by the right and bemoaned by the left.

The decision to privatize the water industry was an apogee of the Conservative government's privatization programme. Water privatization emerged relatively late on the government's agenda; it was not explicitly promoted during Thatcher's first term in office. Undertaken without a clear strategy, water supply privatization was contested within the government and the civil service, as well as by the public (Foster 1992). Despite experience with privatization of network utilities (British Gas had been privatized in 1986, and British Telecom in 1984), the Conservative government made many policy reversals before going ahead with the initiative (Richardson, Maloney, and Rudig 1992).

Part of its hesitation stemmed from the realization that water supply was, somehow, different. Given the nature of the distribution networks, privatized water companies would remain monopolies, at least in the short term. Given the public health and environmental issues associated with water supply, a fairly comprehensive regulatory framework would be required—one that might not mesh easily with the 'light touch' economic regulatory framework Treasury economists had devised to be applied to all privatized network utilities.

Another issue was the sheer size of the capital investment required. The flotation of the water supply industry, with over 50,000 employees and assets valued at over £28bn. (current cost replacement), was one of the largest utility privatizations in Britain to date. Would the capital-intensive industry, facing increasingly stringent EU regulation and an estimated investment requirement of between £24 and £30bn., its assets suffering from years of underinvestment, be able to attract sufficient interest on the part of investors? Most industry observers agree that the companies were deliberately priced low due to Ministers' fears about the failure of water privatization (Richardson, Maloney, and Rudig 1992; Kinnersley 1994); the total value of shares rose by £1bn. to £6.2bn. at the end of the first day of trading (Ernst 1994). The flotation of the water supply companies had succeeded, although questions about underpricing of the original share offer would persist. Although public opinion polls demonstrated that a majority of the population was opposed to privatization (Saunders and Harris 1990, 1994), there was little coordinated public protest, and the original share offers were heavily oversubscribed

(Curwen 1994). The 'British model' was a success, or so it seemed. Few would have predicted that, within a decade, managers of more than one water company would be proposing a return to public ownership of assets.

June 2000: from private to public?

In June 2000, the parent company of Yorkshire Water unveiled plans to radically restructure its water services business: the company would return to public ownership just over ten years after its privatization in 1989. The assets of Yorkshire Water were to be sold to consumers through a non-profit 'community mutual'. While the operation and maintenance of the water supply system would remain in private hands, asset-owning consumers would have direct input into running their local water business.

Customers, promised Yorkshire Water press releases, would benefit from mutualization. New, cheaper financing could be found which would permit increased investment or reductions in bills. The conflict between the shareholder and customer interest would be eliminated, through the alignment of 'ownership of the assets with the interests of consumers, through mutual non-profit based ownership' (Yorkshire Water 2000c). As the company declared in a public statement on the proposal:

> We are proposing this because we think it is in the best interests of all parties—whether customers, shareholders, or the community. Our shareholders, 40,000 of whom live in Yorkshire, get a fair price for the assets in which they have invested in over the last ten years. Our customers get the twin benefits of ownership and in the long term they stand to gain from the cheapest possible way of financing a water service. The community benefits from the protection of vital regional assets in perpetuity. This forever removes the tensions between the interest of shareholders and those of customers. (Yorkshire Water 2000a).

Oddly, given the promised benefits to consumers, there was little apparent interest on the part of Yorkshire Water's customers in the proposal. The company, with 4.5 million customers, received only 15 responses to its proposal. The most high-profile 'public' reaction was that of WaterWatch, a local campaigning NGO that rose to prominence as a watchdog critic of the company's management of the supply system during a severe drought in 1995. WaterWatch strongly opposed the company's proposal; so too did the regulators of the water industry. The debate continued until the economic regulator issued a decisive paper signalling that the mutualization of Yorkshire Water would not be allowed to proceed (Ofwat 2000d). Arguing that the 'present equity-based system has worked well' and asserting that shareholders alone would gain in the short-term from the mutualization, the regulator challenged Yorkshire Water to prove that the mutual business model would be beneficial to customers (Ofwat 2000e). After a high-profile public debate, Yorkshire

Water withdrew its proposal. Yet shortly thereafter another water company, based in Wales, sought and gained permission to convert its core operations into a non-profit corporation owned by its 'members'. Other companies proposed a variety of strategies—in many cases entailing the separation of non-regulated activities from the core, regulated company, and a refinancing of water and sewerage operations with debt rather than equity. The 'British model' of vertically integrated, equity-financed, for-profit provision by private companies had indeed evolved rapidly; and was, some observers argued, under severe strain.

Redefining privatization

The Yorkshire case highlights a long-standing tension in water governance between public and private ownership. As Yorkshire Water argued when promoting its mutualization proposal:

Much of the debate over the water industry in recent years has its origins in public discontent with the concept of privatization of this particular industry: the importance of clean water and efficient sewage disposal to the health and well-being of the community contribute to an intuitive feeling that these assets are more appropriately held in community ownership. (Yorkshire Water 2000*b*)

This 'intuitive feeling' has a conceptual counterpart in arguments made by economists: water supply is characterized by 'market failure', in that it has characteristics (explored in detail in Chapter 2) which prevent markets from functioning efficiently. Many proponents of privatization argue, nonetheless, that water—as with essential goods such as bread or essential services such as electricity—can be supplied more efficiently by the private sector, and that private ownership of water supply infrastructure is not only possible but indeed preferable. Opponents of privatization, in contrast, frequently argue that water supply, as a public good essential for life, ought collectively to be managed in the public interest, whether by the community or by the state. The public/private binary is more than an economic distinction. Here are opposed divergent property forms, management priorities, definitions of equity, and conceptions of how the relationship between state, market, individuals, and the environment should be articulated.

Privatization is thus inherently a political, as well as economic process. For some observers, the government's privatization programme entailed a profound transformation in British political culture and economic life, engendering 'popular capitalism', an 'enterprise culture', and a 'shareholder's democracy' (Pirie 1988). Others depicted privatization as an agenda which merely entrenched class differences and widened wealth disparities (Samson 1994). These interpretations depend in part on assumptions about why and how privatization occurred. Was privatization a strategic manœuvre on the part of Margaret Thatcher and the 'New Right' to redraw Britain's political landscape (Wolfe 1991), or a process of improvisation and persuasion that succeeded only because of the lack of viable financial alternatives to the difficult economic circumstances of the early 1980s (Foster 1992; Clarke 1993)? In many accounts of the British privatization initiative, these two facets of the concept of privatization are invoked (Marsh 1991; Samson 1994). Conventional analyses assume that privatization was either an ideological project, with moral as well as political and economic objectives, or a calculated set of political or economic strategies. Each perspective on this debate captures something important about privatization: it involves a transfer of ownership or control of material assets; it has ideological and political as well as narrowly economic dimensions; and it is related to a broader shift in the 'frontiers of the state'. Yet conventional uses of the term often imply that privatization is a simple act, or contract, and that there is a clear delineation between the 'public' and 'private' spheres, the latter limited to market exchange.

The above interpretations of the term 'privatization' are to some degree misleading, insofar as they imply an overly sharp distinction between 'public' and 'private' management of water. Prior to privatization, private companies supplied water in England and Wales, albeit on a heavily regulated basis. Since privatization, water companies have been treated as private corporations for most purposes, but legally recognized as public bodies for others. In England and Wales, private water companies do not own water outright. Rather, they own water supply infrastructure and hold licences to supply water, granted by the central government, subject to stringent regulation. When using the terms 'public' and 'private', I am referring not to the ownership of water *per se*, but to the ownership of assets (such as reservoirs, network infrastructure, and land) that companies possess, which enable them to supply water. Examples of full privatization are rare in the water supply sector; private sector partnerships are much more common. Moreover, privately owned companies are not always fully commercialized; private firms involved in private sector partnerships in developing countries, for example, frequently operate tariff structures which provide cross subsidies to poorer consumers. The distinction between fully 'public, non-commercialized' and 'private, commercialized' is thus relative, and highly contextual, as water property rights institutions and water regulation norms vary significantly from place to place. The terms privatization and commercialization are thus best interpreted not as referring to a complete, abrupt conversion from monolithic 'public' to 'private' control, but rather as an organizational and/or institutional shift along a continuum of water management options towards the market and private corporations, and away from command-and-control regulation administered by the state. As I argue in subsequent chapters, this should be understood as a process of *re-regulation*, rather than deregulation, in which the state (and other actors) engages in institutional learning. This process of re-regulation may entail an

The environmental dimensions of privatization

Privatization is often analysed in terms of its economic, and also its political dimensions. The political agendas of privatization and regulation have been the focus of much of the literature produced about the water supply industry since 1989 (see, for example, Maloney and Richardson 1994; 1995). Much research has been conducted from the perspective of regulatory economics—which asks questions about efficiency, performance, and incentives (see, for example, Hunt and Lynk 1995; Saal and Parker 2001). Other research—broadly sociological—has focused on labour issues (O'Connell-Davidson 1993; Mulholland 2002) and consumer issues (Drakeford 1997; Ernst 1994; Graham and Marvin 1994; Saunders and Harris 1990). Relatively few studies, in comparison, have been conducted of the environmental dimensions of privatization. In contrast, the environmental impacts of privatization, and also the environmental drivers of privatization are a focus of this book. At a conceptual level, this entails an attempt to understand the mutual interrelationship between socio-economic restructuring and environmental change. At an empirical level, this entails integrating environmental variables into political economic analysis—simultaneously addressing the social, political, economic, and environmental dimensions of privatization.

This approach is predicated upon the understanding that water regulation in England and Wales (as in many other places) has an increasingly 'green' character. Over the past two decades, environmental issues have been formally integrated into water resources planning, an environmental regulator has been created, and the water industry has to some degree reinvented itself as an 'environmental services' industry. Water quality and environmental expenditure are key drivers of capital expenditure programmes in the industry. At the same time, water companies have adopted market-derived principles—such as economic efficiency—into management practices. Environmental regulatory framework has similarly adopted techniques—such as cost-benefit analysis—derived from the private sector. Water companies in England and Wales are much more environmentally sensitive, and to a much greater extent guided and constrained by environmental regulations, than they were two decades ago.

From 'state hydraulic' to 'market environmentalist' regulation

To characterize this emerging approach to water supply management, I use the term 'market environmentalism'. This term attempts to capture the generalized and profound change ongoing in our relationship with water—the move away from state towards market mechanisms in resource policy-making, regulation, allocation, and ownership. Market environmentalism entails the pursuit of environmental ends through market means (Anderson and Leal 1991). In the case of the water sector, market environmentalism entails an emphasis on efficiency, economic equity, planning for scarcity, demand-led solutions, regulatory decision-making based on market-simulating and surrogate valuation techniques, an emphasis on incorporating environmental goals and valuations into water policy, and private management and/or ownership of infrastructure. This contrasts with the 'state hydraulic' paradigm of water regulation, dominant in Britain (and in many countries) throughout much of the twentieth century, which entailed a focus on social equity and universal provision, planning for growth, supply-led solutions, command-and-control regulation, a discursive representation of nature as a 'resource', and state ownership. Much of the book is devoted to exploring why the former regulatory approach supplanted the latter (particularly Chapters 2 and 3), and to linking the environmental impacts of privatization to the contested politics of water supply management which have repeatedly pitted environmental concerns against consumer concerns in post-privatization debates.

The politics of environmental management in the post-privatization water industry must be understood with reference to ongoing attempts to commercialize water supply. Commercialization, as originally envisioned by the architects of privatization, would imply the introduction of competition and cost-reflective pricing, incorporating externalities, and relates prices to costs imposed by individuals on the water supply system; requiring a network or other mode of supplying water in standardized units; and volumetric metering of all consumers as a means of communicating price 'signals'. These changes have proved difficult to introduce in the case of networked water supply; as a result, the commercialization of the water supply industry has been relatively limited when compared to other networked utilities.

Chapter 2 attempts to explain why commercialization in the water supply sector has been relatively limited, rooting the explanation in a characterization of water as an 'uncooperative' commodity, making reference to neoclassical and political economic interpretations of 'market failure'. A typology of water supply privatization is presented, and a genealogy of ideas about water as a resource are analysed as a means of exploring the key economic, political and environmental arguments made in favour of water privatization and commercialization. This requires a dialogue between regulatory and natural resource economics and various strands of what may broadly be termed a political economy approach, revisiting definitions of regulation and commodities. In this reading, privatization is understood to be embedded in a longer-term and generalized transformation in our regulation of nature. This presupposes neither that privatization was a distinct rupture with the past, nor that it simply represents 'natural' and logical continuities with past practice; both are true to some degree. The periodization of history implied by this analysis is characterized as the gradual, partial evolution away from a 'state hydraulic' to an emerging 'market environmentalist' mode of regulation.

In order to explain the emergence of the state hydraulic mode of regulation, Chapter 3 returns to the nineteenth century, a critical period in the industrialization of water supply, during which extensive hydraulic networks were built. The chapter focuses on the creation of hydraulic networks and the evolution in frameworks for regulation of these networks in the context of urbanization and industrialization, beginning with the shift from private management to municipalization in the mid-1800s through to nationalization in 1974 and privatization in 1989. As networks accrete over time and across space they simultaneously embody historical as well as contemporary management practices. As such, infrastructure networks enable continued production, but simultaneously pose barriers to new modes of regulation insofar as they 'fix' production at certain scales, and abstraction and distribution in specific patterns. The tension between historical and contemporary practices is particularly acute in the case of water supply, given the typically long life span of its infrastructure and high 'sunk' costs associated with network renewal or replacement. Certain characteristics of the water network—for example, the widespread absence of metering—reflect policy goals of the previous, nationalized era rather than the current regulatory regime. Post-privatization, for the majority of consumers in England and Wales, charges for water services continue to be based on the rateable property value of their residence, not on the amount of water consumed. In the absence of volumetric charging and competition (as potable water is confined to networks that have become naturalized monopolies), the prices which consumers pay for their water services are determined not by supply and demand, but by a complex series of regulatory interventions designed to simulate the market. This is one example of how the analysis of the creation of the water supply network in the late nineteenth and early twentieth centuries facilitates understanding of the constraints that have impeded the full commercialization of the water industry.

The emergence of market environmentalism in the water industry is the focus of Chapter 4, which focuses on the commercialization of the water supply industry. The chapter examines three aspects of commercialization: valuation; the creation of price signals (via metering, with the goal of facilitating allocation of water to its highest value uses); and liberalization (the introduction of competition). The chapter argues that each presents trade-offs, implying dilemmas for regulators attempting to introduce a greater degree of commercialization into the water supply industry. To some degree, the government has responded to these trade-offs by retreating from its earlier commitment to full commercialization of the water industry; the analysis of reasons for and implications of this 'retreat from commercialization' is begun here, and continued in subsequent chapters.

A central focus of Chapter 4 is the evolving definition and role of efficiency; the chapter does not analyse improvements in productive efficiency, but focuses instead on integration of efficiency into environmental management. The question of efficiency deserves careful attention not only because it lies at the heart of the regulatory framework, but also because greater efficiency was one of the central promises of privatization. An underlying justification for privatization was the belief that the costs of 'state failure' are greater than that of market failure (Martin and Parker 1997). The state, it is argued, is a less efficient provider of public services than the market. Water companies would thus, it was argued, operate more efficiently in a market environment, albeit where market forces are simulated by regulatory assessment of comparative performance (Littlechild 1988; Vickers and Yarrow 1988; Martin and Parker 1997).

The implications for consumers, the environment, and the water supply industry

The degree to which the concept of 'state failure' and the search for increased productive efficiency was a motivation for privatization continues to be a matter of debate. Some observers would prioritize short-term political motivations, and argue that privatization was pursued 'for political reasons related to the government's troubled attempt to manage the economy and stay in power, rather than to the economic pursuit of efficiency in the industries concerned' (Clarke 1993: 207).

Leaving the debate over the motivations of privatization aside, the economic arguments in favour of privatization were pivotal in shaping the post-privatization regulatory framework, particularly in the priority given to efficiency measures. Privatization was recognized to be an insufficient condition for the 'optimization' of water management—defined, in distinct contrast to the past, as the maximization of efficiency. A complex set of regulations and series of regulatory interventions were thus designed at the time of privatization to permit the privatized industry to function, and to enable the continued commercialization of the industry. Part II is devoted to an examination of the evolution of this regulatory framework, which has entailed an expansion of state intervention and regulatory oversight in the water sector; successive chapters are structured around the mechanisms and implications of re-regulation for consumers, environmental management, and the water supply industry.

Chapter 5 examines the implications of the Yorkshire drought of 1995: the most extreme climate event and the most negative public relations episode faced by the water companies since their privatization in 1989—for consumers and the environment. The management of the drought by the regional private water supply company sparked a debate concerning the ability of private companies to provide sufficient supply security for consumers. In this chapter, the drought is interpreted as a 'revelatory crisis', unveiling disincentives in the regulatory framework leading to a lack of attention to, and underinvestment in supply system security. The supply system crisis is analysed as simultaneously naturally, socially, and discursively produced through a complex admixture of meteorological conditions, regulatory decisions, and company and consumer actions. The chapter outlines the series of regulatory interventions which followed the drought, as regulators and government realized that the privatized framework did not necessarily provide sufficient incentives for water conservation or against 'environmental raiding' during droughts.

Chapter 6 examines the implications of the progressive commercialization of water pricing policy—arguably the most controversial area of water policy—for consumers. The analysis attempts to demonstrate the concrete implications of the shift in policy from the principle of social equity to economic equity in water pricing. The chapter provides a genealogy of concepts of efficiency and equity in water regulation over the past thirty years, prior to evaluating the implications for water consumers of one aspect of this process—the shift away from policies prioritizing inter- and intraregional equalization (implying a principle of social equity) towards policies prioritizing economic efficiency (implying a principle of economic equity) in water charging. The shift from equalization to (neoclassically defined) equity, predicated on an assumption of water as a scarce resource, has resulted in increasing the spatial and social differentiation of costs of water supply and services.

The perception that the regulatory framework was inadequate for protecting public health and certain environmental goods led to a re-prioritization of social and environmental policy goals within the water policy framework. *Ad hoc* regulatory intervention and the imposition of command-and-control regulatory techniques were designed to offset market gains for shareholders; mitigation of market abuses, rather than maximization of the operation of market forces, was the dominant character of regulatory intervention at the end of the first decade following privatization. As outlined in this chapter, re-regulation of the water industry was provoked by consumer dissatisfaction with the distributional impacts of the water pricing regime. This re-regulation has in turn been a driver for the ongoing wave of restructuring of the water supply industry, which is examined in the following chapter.

Chapter 7 examines the ongoing restructuring of the water industry, reviewing companies' strategies of vertical de-integration, diversification, and internationalization. The analysis then turns to a consideration of the future of the English and Welsh water companies and the regulatory framework of price cap regulation of privatized utilities. The failure to create a regulatory framework that adequately addresses the equity and externality dimensions of water supply has resulted in a substantial re-regulation, and significant alterations to the English framework of price cap regulation as a regulatory and financial model. As a result, the companies created at the time of privatization are increasingly seeking to free themselves from the constraints, and reduced profitability, of the progressively more stringently regulated domestic water supply business. The chapter examines three strategies pursued by the companies: diversification; internationalization; and vertical de-integration.

This restructuring, combined with the progressive re-regulation of the industry, has called into question the long-term stability—and perhaps viability—of the British model of water privatization.

Chapter 8 concludes with a set of reflections on questions likely to be of interest to those involved in debates around water supply privatization, whether in Britain or abroad. The lessons that might be drawn from the experience with water privatization in England and Wales are discussed. The robustness of the British model, and the likelihood and appropriateness of its export overseas are explored. The findings of the book, in terms of the implications of the British model for equity, access, and environmental management are summarized. The chapter also addresses the question of the significance of the recent trend towards restructuring (including vertical de-integration, refinancing, and mutualization) in the water supply industry. Restructuring within the industry has entailed a move away from the original 'British model' of vertically integrated, equity-financed utilities; the chapter speculates about what the implications might be, and asks what the future might hold for the British water supply industry.

2

Water: An Uncooperative Commodity

During the past century, in many countries, water management was characterized by the dominant role of the state as owner, manager, and regulator of water supply infrastructure. Yet state dominance of the water supply sector is relatively recent. Prior to the mid-nineteenth century, state involvement in water supply was limited; networked water supply, where it existed, was largely the preserve of private companies, or self-provisioning communities. The predominant role of the state in water supply management is, in most industrialized countries, a relatively recent historical development. How, then, do we explain the growth of public sector control, the resurgence of private involvement in the water sector, and the re-emergence of the 'state versus market' debate at the start of the twenty-first century?

This chapter presents a conceptual framework for understanding how and why water supply has shifted between private corporate and state control over the past two centuries.[1] Privatization is thus positioned in historical context. In explaining why and how water supply is regulated, and why modes of regulation change over time, the chapter adopts a political ecological framework—understood as the 'political economy of socio-environmental change'. In analytical terms, a political ecological approach emphasizes the interrelationships between material (physical), socio-economic, and representational (discursive) aspects of resource regulation. This chapter attempts, then, to trace the relationship between socio-economic restructuring and environmental change in the water sector. This approach is thus somewhat distinct from analyses presented in the water management, political economy, and regulatory economics literature, although it engages these in debate.

In laying out the conceptual framework, three points are emphasized. First, a substantive change is occurring in our relationship to, and management of water supplies. I have characterized this change as a multi-dimensional shift from a 'state hydraulic' to a 'market environmentalist' paradigm of water management. The first part of the chapter explores these two paradigms, and

---

[1] This argument was developed through my analysis of the English and Welsh case. To some degree, it will be applicable to other industrialized societies in the 'North', but countries in the 'South' present a different, and probably more complex dynamic.

analyses the some of the concepts which underlie them; most importantly, concepts of 'market failure' and 'state failure' respectively. Second, privatization is re-conceptualized as a process of re-regulation—which does not necessarily entail deregulation. A portion of the chapter is devoted to exploring the different definitions and processes of re-regulation. Third, privatization of water supply, in particular, is likely to entail re-regulation, thereby increasing the oversight functions and interventionist role of the state. This is in part because the application of 'market environmentalism' to water is problematic (and perhaps ultimately unsuccessful), given water's biophysical characteristics, which make it difficult to commercialize and privatize. In analysing these biophysical characteristics (or the 'materiality' of water), I characterize water as an inherently 'uncooperative' commodity, and argue that the privatization and commercialization of water is inevitably fraught with difficulty. This chapter tries to explain why this should be so, at a conceptual level. Subsequent chapters explore whether this has indeed been the case in England and Wales, and attempt to outline some of the implications for consumers, the environment, and the water supply industry.

Market failure and the 'state hydraulic paradigm'

Throughout much of the twentieth century, water was mobilized as a strategic resource for societies undergoing modernization, industrialization, urbanization, and agricultural intensification. A prevailing assumption amongst water planners was that sustained growth in water use was necessary for economic growth; and that sustained growth in water demands was inevitable, given economic growth, increasing population, and increasing consumption per capita. Emphasis was accordingly laid on developing new sources of supply, often to support specific industrial sectors.

State-supported hydraulic development mobilized water resources in vast quantities. In the five decades following WWII, the number of large dams (defined by the International Commission on Large Dams as being over 15 metres in height, or between 5 and 15 metres with a reservoir storage of more than 3 million m^3) has increased from approximately 5,000 to over 45,000 (McCully 1996; World Commission on Dams 2001). Water resources development was critical to the industrialization of agricultural production, and to the implementation of the 'Green Revolution' in many countries; globally, irrigated area has increased from approximately 50 million hectares to 300 million hectares over the twentieth century (Gleick 2000). In many catchments, a significant proportion of renewable resources have been harnessed; in the basin of the River Thames, for example, approximately 55% of effective rainfall (a proxy for renewable resources) is annually abstracted for human use (NRA 1994*a*).

In many European countries, the state developed water resources as factors of production, treating water as a strategic resource (like steel or oil); an essential input to the economy, and thus an imperative for large-scale infrastructure projects to secure supplies. The level of investment was significant. In the United States, capital investment for water over the twentieth century is estimated at US$400m. (unnormalized), the bulk of this having been spent on large-scale engineering projects (Rogers 1993). In Spain—which has one of the largest proportions of surface area covered by reservoirs in the world—the nominal economic value of the water resources made available through hydraulic development has been calculated at a range of between 5% and 8% of GDP (Martín Mendiluce 1996; Ministerio de Medio Ambiente 1998). Given the scale of investment and long lead times, private sector provision (particularly of water supply if universal coverage was desired) was thought to be unfeasible, and hence state financing of water infrastructure (particularly for water resources development) assumed to be necessary. In various forms, then, particularly in the post-war period in the twentieth century, water management, and investment in the water sector by the state, played a supportive role in capital accumulation as a vital input to processes of urbanization, industrialization, and agricultural modernization while acting as an important mechanism of social and political legitimization (and, in some cases, territorial colonization) by the state.

Water management practices in industrialized countries in the mid-twentieth century varied considerably, of course, but were often characterized by a high degree of state management and ownership, and state subsidization of infrastructure development and, in some cases, operating costs. Regulatory control, in most OECD countries, took the form of state-run regulatory frameworks and, often, public ownership of water supply infrastructure. Drawing on Moral and Saurí (2000), this mode of management can be termed the 'state hydraulic' paradigm, which is characterized by: planning for growth and supply-led solutions, with an emphasis on hydraulic development as a means of satisfying water demands; a focus on social equity and universal provision; command-and-control regulation; a discursive representation of nature as a 'resource'; state ownership and/or strict regulation of water resources development and water supply provision; based on a desire to provide sufficient quantities of water, where and when needed, such that economic growth could proceed unconstrained.

Water supply: a material emblem of citizenship

The principles of state hydraulic management were applied to water for human consumption ('drinking water' or 'water supply') in different ways in different jurisdictions, although the provisions for 'universal service obligation' and cross-subsidies were widespread. In England and Wales, the entire water use and waste water disposal cycle was brought under municipal and later centralized public ownership. In France, private companies operated as service providers to infrastructure-owning municipalities. In Spain, investment in water resources was crucial to Franco's project of agricultural modernization, and the state assumed control over surface water resources across all sectors (Bakker 2002). The United States followed a more decentralized model, with local providers—the majority of which remain municipally owned—operating on a largely local basis. Notwithstanding these variations, in most industrialized countries, the majority of water supply infrastructure by the early twentieth century was owned by governments, in contrast to the earlier predominance of the private sector in networked water-supply provision.

Municipal ownership, in particular, was associated with the extension of network water supply provision. In the mid-nineteenth century, for example, less than 10% of British households had a piped-water connection (Gregg 1950); by the mid-twentieth century, over 90% of England's population was connected to a networked water supply system (Sleeman 1953). This rapid extension of drinking water supplies was often funded largely by local governments, partly through local taxes and partly through subsidies (in the form of direct transfers or financing with long repayment periods, low-interest and/or high discount rates), usually from higher levels of government. The state-led expansion of drinking water supplies was underpinned by a model of social welfare in which state provision of public and merit goods was thought to be in the general economic and social interest (Ernst 1994; Graham 1997). Drinking water supply was conceived, in many instances, as a welfare service, with critical impacts on public health and environmental quality. The state bore, in large part, the costs of drinking water supply and resources development. This behaviour was not unusual at the time; the implementation of the state hydraulic mode of regulation occurred over the twentieth century (at least in most OECD countries), and particularly post-WW II, during which states undertook to provide those services assumed to be unprofitable or unfeasible for the private sector.

As with other welfare services, non-market norms and rules applied to the provision of water supply. In Britain, cross subsidies between classes of consumers were implemented through linking water bills to property values rather than metering domestic supplies. Similar policies were applied in other countries. In Spain, for example, underpricing of water in the agricultural sector facilitated the agricultural development which was central to Franco's autarkic political economic project. These policies might be narrowly interpreted as a class accommodation, with the state mediating the inevitable conflicts over water use and facilitating cross subsidies from one class of users to another. Yet more than merely socio-economic logic applied to water resources. Drinking water supply, in particular, fell into what Walzer would term a separate 'sphere of justice' of goods—essential for dignity and for participation in public life—created by the welfare state. In most industrialized countries, sufficient supplies of potable drinking water came to be understood as an entitlement to which all citizens had access (not the case, of course, in all countries). This 'right' was not formalized. South Africa is the sole country whose constitution guarantees its citizens a right to water.[2] Rather, drinking water, as with other welfare services, was included in a broader set of socio-cultural entitlements to which all citizens had access; material 'emblems' of citizenship, which were the moral duty of society, via the welfare state, to provide.

Market failure

The state hydraulic paradigm was frequently justified through reference to the 'market failures' which characterize water supply provision: instances where water refuses to cooperate with the standard behaviour of commodities. There are several market failures usually identified with respect to water supply: monopoly; externalities; and public goods (Cowan 1993; 1997*b*). Water supply is cited as an example of what economists term a 'natural monopoly', insofar as supply by one firm entails lower costs than supply by more than one firm; the incumbent firm has an overwhelming cost advantage as compared with new entrants. Natural monopoly arises in the water industry in part because 'direct competition between firms in the provision of [infrastructure] networks . . . would entail inefficient duplication of fixed assets' (Cowan 1993: 14). The incumbent firm will have an overwhelming cost advantage with respect to any new entrant, and thus the market will tend to be characterized by only one seller, and hence prey to the problems associated with monopoly power.

A second type of market failure identified by economists is that of 'externalities': costs or benefits arising from water production not accounted for in the price mechanism, which thus do not accrue to the producer. These externalities may be both negative (for example pollution) and positive (for example the social benefits gained from the widespread availability of safe and adequate water supply and sanitation) (Winpenny 1994). It is important to note that externalities are understood to be, to varying degrees, intrinsic to the production of water, regardless of the broader economic system. The argument is not that capitalism produces externalities; rather, economists observe that the conditions for the robust functioning of a market are not observed with respect to water supply, insofar as externalities are not incorporated in the price of water. Given these negative externalities, proponents of the 'market failure' argument typically characterize water supply, or at least the public health outcomes associated with water supply infrastructure, as a 'public good' (non-excludable and non-rivalrous).

From this perspective, networked water supply lies within a sphere of goods and services for which the market, left to its own devices, does not function well; it deviates from the standard behaviour of commodities—hence its 'fail-

---

[2] The Constitution of the Republic of South Africa guarantees the right of citizens of access to sufficient water (Act 108 of 1996, section 7(2)).

ure'. Indeed, the health and hygiene effects of lack of access to water, together with the tendency of private companies to fail to extend coverage to the poor (both as a result of the tendency to cherry-pick profitable neighbourhoods and classes of consumers, and the high prices and poor services resulting in a situation of natural monopoly), were two of the most important justifications for bringing water supply under the control of the state, through the creation of regulatory oversight mechanisms. In addition to the 'market failure' argument, the symbolic and cultural importance of water as a (partially) non-substitutable resource essential for life, its strategic political and territorial importance, the intense conflicts that arise over the use of a flow resource required to fulfil multiple functions (agricultural, industrial, drinking water, environmental), and the need in industrialized, urbanized societies to mobilize large volumes—invariably at a high cost relative to the economic value generated, implying large, long-term capital investment requirements which private companies were not always willing to assume—have been used, particularly in the twentieth century, to justify public-sector investment and ownership.

State failure and the 'market environmentalist' paradigm

By the final quarter of the twentieth century, the costs of the state hydraulic paradigm were receiving greater attention. At the pinnacle of the state hydraulic paradigm stood the dam-as-icon; but cost-effective dam sites had become more difficult to find in most developed countries, and their social and environmental costs were becoming more apparent. After half a century of experimentation, broad-based questioning of the cost-effectiveness of hydro-development undermined the assumptions (life-span, hectares irrigated, consumers supplied) underlying the often optimistic forecasts of water resources developers (see, for example, World Commission on Dams 2001). Undermining the growth–supply nexus justifying continued expansion of water resources infrastructure was a stagnation of demand in many industrialized countries. In England, water demand declined in 1980 (excluding drought years) for the first time since record-keeping began (WSA 1987). In the United States, water withdrawals began declining in the mid-1980s and are now 10% below their peak (Gleick 2000). 'White elephant' reservoirs further undermined the case for supply-led management strategies and new hydro-development. Premised on forecasts in the early 1970s of rapid growth in industrial demand in the Newcastle area, the Kielder reservoir in north-east England was completed in 1982, by which time demand had dropped so significantly that there was 100% spare capacity in the local water supply network (Pearce 1982). Over the same decade, years of lobbying by conservationists and environmental groups began to change public consciousness. Greater awareness emerged about the effects (and often unquantified costs) of hydraulic development: extirpation of aquatic species (particularly fish); displacement of communities; flooding of cultural sites; contamination of water sources; disruption of ecological processes (such as sedimentation); and environmental degradation (Gleick 2000).

The state hydraulic paradigm thus faced a multidimensional challenge: environmental, social, and economic. In a period of deep fiscal constraints, with governments less willing to subsidize infrastructure development, public environmental concern combined with technological innovations to increase the appeal of alternative approaches to water management. In many countries, considerations of water conservation began to be taken more seriously in water use planning debates. In England, a turning point was reached in 1981, when the government rejected the Ennerdale scheme (converting a lake into a reservoir in the touristy Lake District) on the basis of insufficient water conservation measures by the local water company (Pearce 1982).

From market failure to state failure

The challenge to the state hydraulic paradigm must be understood within the context of a more generalized critique of the role of the state, and particularly the welfare state, which came to the fore of political debate from the 1970s onwards. Critics from both the left and the right seized upon the slackening pace of growth of social expenditures in the post-'oil shock' as evidence of a crisis of the welfare state (Pierson 1998). As Pierson notes, perspectives on the root causes of the 'crisis' diverged considerably. From the Marxist perspective, the fiscal crisis experienced by many welfare states arose from the contradiction between the capital accumulation supporting and political legitimation functions of the state (O'Connor 1973). From the neo-liberal perspective (presented in its most sophisticated form by public choice theorists), the conflicting incentives arising from short-term political incentives and economic imperatives inherent in a political system of liberal democracy (and politicians' tendency to resolve this dilemma through economically unsound decisions) were the root cause of the fiscal crisis of the state (see, for example, Olson 1965; 1982; Tullock 1976).

It is important to note that disagreement continues not only about the causes of states' fiscal crises in the 1970s and 1980s, but also about their very existence and severity (Pierson 1998). For our purposes, the answer to the question of whether fiscal limitations were externally imposed or arose from the 'wilful incompetence of the state' is less important than the discursive role played by fiscal crisis at the macroeconomic level, which in turn supported underinvestment at the sectoral level. In many countries where a high degree of control was assumed by the state over water resources, continued public provision was, by the end of the twentieth century, being undermined by the contradiction which beset public services provision more generally: the continued legitimacy of the state was dependent upon the satisfaction of expectations that it had itself sanctioned, but which threatened to undermine either environmental sustainability and/or economic competitiveness (Hay 1996). In the case of the water sector, states frequently justified underinvestment in infrastructure and services through reference to the lack of public finance. Underinvestment in water infrastructure was a politically expedient strategy, insofar as impacts of under-investment may take much longer to be evident to the public than in other sectors, such as health care. The resulting decline in service provision standards (declining quality or quantity, rising prices) undermined the legitimacy of the state as service provider, in turn providing justification for privatization and commercialization. The crisis of the 'administered' mode of regulation experienced by some states in the final quarter of the century thus both contributed to, and was exacerbated by the breakdown of the state hydraulic paradigm.

Regardless of the veracity and cause of the 'fiscal crisis of the state', there is general agreement on the broad contours of the response. During the final quarter of the twentieth century, the concept of 'state failure' gradually displaced the concept of 'market failure' which formerly underpinned arguments in favour of public ownership and management of many public services. According to the state failure hypothesis, states are unproductive, inefficient, and ineffective (and from some neo-conservative perspectives, despotic, and inimical to freedom) (Pierson 1998). An underlying assumption of the state failure hypothesis is that the market is more efficient than government at providing basic services. In developing countries, the ensuing mobilization of the structural adjustment of states is commonly referred to as the 'Washington Consensus', characterized throughout the 1980s and 1990s by core programmes of fiscal austerity, privatization, and market liberalization (Stiglitz 2002), often conjoined with free trade, financial liberalization and foreign investment incentives, low marginal rates tax, high real interest rates, and flexible labour markets (Bond 2001). In most industrialized liberal democracies with welfare states, an equivalent (although often self-administered) 'structural adjustment' occurred in the 1980s, often under the guise of the 'New Public Management' (Osborne and Gaebler 1993; Pierson 1998; Walsh 1995). Like the Washington Consensus, the New Public Management is rooted in the conviction that the private sector is more efficient and effective than public bureaucracy. The preferred solution is deregulation, and/or the transfer of functions from the state to the private sector through privatization, where possible. Where privatization is not possible, proponents of the New Public Management advocate the application of (ideal-type) principles of business management to public sector functions; in other words, the commercialization of public administration.

Market environmentalism and water

The emergence of market environmentalism in water management debates was premised, in part, on the state failure hypothesis. Whereas the concept of market failure once underpinned arguments in favour of state provision, state failure was increasingly invoked, from the 1980s onwards, to argue in favour of privatization and commercialization of water supply. Proponents of the involvement of the private sector argued that this would lead to an increase in efficiency, in part due to the inherent ability of the private sector to innovate, and partly through removing social policy goals—such as employment generation, or (more commonly) wealth redistribution through cross-subsidy—from water policy.

Given state failure, argued the most important multilateral lender for the water sector, 'there is no good economic reason for state ownership to persist in tradable-goods industries' (World Bank 1997: 64). Underpinning this argument is a dual discursive move. The category of tradable goods has been expanded to include water: no longer a public good subject to market failures which must be supplied by the state at subsidized prices, but a tradable good which can profitably be supplied by the market under competitive conditions. The utility sector is simultaneously re-conceptualized as potentially profitable (running counter to assumptions held throughout much of the twentieth century), rather than a provider of strategic resources in need of subsidies. Citizenship is reconfigured: if water is a tradable good rather than a public good, it follows that consumers of water are categorized as customers rather than voters, who have access to water through their purchase of water as a commodity, rather than the right as citizens to a water supply service. Water supply provision is similarly reconfigured as a business, rather than a public service, which (whether under public or private ownership) should have as its primary goal the maximization of efficiency rather than social equity, in the context of the increasing scarcity of water resources.

Two dimensions of efficiency are important here: water conservation (greater efficiency in water use); and water use (more efficient allocation of water between different value uses). The organizational shift from state to market, a change in management strategies from supply development to conservation, and a perceptual shift from water 'abundance' to water 'scarcity' are thus often intertwined in current water policy debates. The logic of the market implies the need for, and introduction of techniques which prioritize efficiency, which is in turn presumed to facilitate conservation; hence the strategic alliances which frequently form between ecologists and economists in support of market environmentalism. It bears noting that the 'environmental' movement is far from homogeneous, and environmentalist concerns play an important role in questioning the 'state hydraulic' paradigm from a variety of perspectives. In the environmental movement, however, the ethic of care underlying earlier concerns with conservation and preservation has been to a degree displaced by a discourse of sustainability and associated logic of compensation, enabling convergence with economists in advocating an ethic of efficiency in resource management.

This integration of the goals expressed in the Washington Consensus and the New Public Management, and that of the environmental movement, is not unique to the water sector. Market environmentalism can be viewed as one expression of ecological modernization—a term which both describes a philosophical approach to environmental reform, and a programme for the environmentally driven reform of business. Eco-modernization theorists view the environmental crisis as creating opportunities rather than posing threats, and seek to move beyond apocalyptic viewpoints, emphasizing the potential for innovation and transformation of science, knowledge, and modernity itself (Mol and Sonnenfeld 2000). As a programme of action for business (see, for example, Hawken and Lovins 1999), ecological modernization emphasizes: science and technology as a source of innovative responses to, rather than a cause of environmental problems; the increasing importance of market dynamics and economic agents (and decreasing importance of the state); the transformations in the role of the nation-state (from government to governance); modifications in the position, role, and ideology of (particularly environmental) social movements; changing discursive practices and emerging new ideologies (such as an increasing acceptance of the principle of intergenerational equity). As a manifesto for the 'greening of capitalism', then, ecological modernization offers the hope of a virtuous fusion of economic equity and efficiency (and hence growth) with environmental conservation (Christoff 1996; Mol 2002; Hajer 1995; 1996).

Market environmentalism in the water sector, as with ecological modernist approaches more generally, promises environmental ends via market means. Frequently, it entails an emphasis on planning for scarcity, demand management, regulatory decision-making based on market-simulating and surrogate valuation techniques, and a growing emphasis on incorporating environmental values into water management decision-making, and private management and/or ownership of infrastructure (Table 2.1). Economic efficiency is prioritized, while dispensing with the assumption of the link between economic growth and water use. Water conservation is balanced with, and in some cases prioritized over security of supply. Technically, the focus shifts from creating new sources of supply to managing demand—through a variety of techniques (conservation, new low water use technologies, alternative supplies (grey water, reclaimed waste water, desalinated water, recycled water), metering, new tariff structures, and educating consumers in a new 'ethic' of water use). Economic equity (the 'benefit' or 'willingness-to-pay' principle) displaces social equity (the 'ability-to-pay' principle) in water pricing. Consumer access is legitimated not by a citizen's entitlement to water as a service, but by a customer's purchase of water as a commodity. A reconfiguration of the hydrosocial contract between users and their environment is required; consumers paying per unit volume at cost-reflective prices will use water more efficiently than unmetered households or farmers accustomed to treating water as a public service. This argument is supported by a discursive repositioning of the concept of water availability: water scarcity is depicted as a universal condition—simultaneously natural, justifying a new ethic of efficiency and the

Table 2.1. Water management: state hydraulic versus market environmentalist paradigms

	State hydraulic	Market environmentalist
Economic regulation	Command-and-control	Market-based instruments
Resource management	Growth-oriented, supply-driven	Scarcity-responsive, demand-led
Network manager	State	Market
Primary goals	Universal provision; quantity	Efficiency; quality
Provision ethos	Service	Business
Consumer identity	User	Customer
Method of charging	Unmetered	Metered
Raw water (pricing)	Resource (subsidized or free)	Environment (abstraction priced)
Water supply pricing	Social equity (ability to pay)	Economic equity (benefit principle)

commercialization of water, and social, the result of flawed public management, justifying the privatization of water. This serves as a further justification for water commercialization: if water is an increasingly scarce resource, it requires efficient management, which (if we accept the claim of state failure) only the private sector can provide.

An uncooperative commodity

The claim of generalized water scarcity deserves close scrutiny. Water is a resource mobilized by humans on a massive scale. Humans withdraw 5,200 cubic kilometres—or 5.2 trillion metric tons—of water annually (approximately 10% of total surface run-off) (Gleick 1993). These global figures mask, of course, the high degree of spatial and temporal availability of water; withdrawals are greater than 50% of run-off in some regions (North Africa, Central Asia, South-Western US, South-Eastern England). In these regions, water quantity, and in more humid regions, declining availability due to declining water quality, are the causes of the scarcity experienced by humans. Scarcity, in other words, is socially produced (sometimes termed 'second-order' scarcity). That our awareness of scarcity is growing is a signal not of *absolute* scarcity, but of *relative* scarcity, due to factors such as increasing pollution, population density, and water use per capita.

Within the constraints of the hydrological cycle, the amount of H_2O is limited, insofar as there is a finite amount of freshwater available within the biosphere at any one time. Acknowledging this finite limit tells us little, however, about human use of natural resources, which can only be relationally defined with respect to human needs, practices, institutions, and technologies (Harvey 1996). Harvey offers a useful definition of a natural resource that captures the mutually constitutive relationship between nature and humans: 'a cultural, technical and economic appraisal of elements and processes in nature that can be applied to fulfil social objectives and goals through specific material practices' (Harvey 1996: 147). Scarcity of water as a resource is thus defined in relation not only to human needs, but also to specific practices of provision and use of water. Whether or not water is *scarce* depends on such varied factors as population density and distribution, sanitary habits, water distribution systems, and customary leisure and amenity uses; scarcity is dependent on the 'hydrosocial', in addition to the hydrological, cycle. The scarcity of water, as of any other resource, thus 'presupposes certain social ends, and it is these that define scarcity just as much as the lack of natural means to accomplish these ends' (Harvey 1979: 178).

The intensification of nature accumulation

From a political economic perspective, then, externalities should be interpreted not as a symptom of failure to comply with an idealized market model, but rather as an inevitable by-product of the production of drinking water, and potentially an opportunity, as well as a cost, for capital. Capital may take advantage of externalities by internalizing them, investing in their mitigation or eradication as a service, or thereby producing 'goods' for which it may charge. Companies which sell technologies for or engage directly in pollution clean-up activities are a classic example of the profit potential in the internationalization of negative externalities. The more general observation follows that threats to the material and human bases of reproduction (for example, declining water quality) are both a threat to, and an opportunity for, capitalism (Enzensberger 1996).

In some cases, 'the commodification of nature is capital's (political) response to ecological degradation that has become a barrier to further capital accumulation' (Benton 1996: 192). Both capital and the state respond to these barriers in a variety of diverse, creative, and constantly evolving ways: capital seeking profit; the state seeking to develop a mutually supportive relationship between capital accumulation and regulation, enabling economic growth and creating the conditions for political stability. Evading barriers to capital accumulation—devolving the costs of clean-up of environmental 'externalities' on to others, or moving from one geographical region to the next in search of ever-cheaper costs or readily available factors of production—is a long-established strategy of capital. Subsidizing capital accumulation through assuming responsibility for environmental protection and rehabilitation is a long-established strategy of the state. Yet these strategies are becoming less viable given what Katz (1998) terms the 'involution' of nature at the end of the twentieth century, in which extensive accumulation strategies and the associated instrumentalist approach to nature as a 'resource base', are reaching (human-perceived) limits. As Altvater has noted, extensive accumulation entails a discounting of the future, which routinely leads to overexploitation, whose costs are often hidden, because 'unvalued' (Altvater 1993). As a result, the growing economic, political, and ecological costs of these naturally mediated unintended consequences of production have become significant, and widely recognized.

In response has emerged, over the past two decades, a generalized transformation in modes of resource accumulation, across resource sectors and locales (Castree and Braun 1998; Escobar 1996; Katz 1998; O'Connor 1996). From a political ecological perspective, we are witnessing a qualitative transformation in strategies of nature accumulation, in which the expansion of capital is increasingly intensive as well as extensive[3] (Table 2.2). In Martin O'Connor's analysis, the intensification of nature accumulation is a response within capitalism to the '(i) ostensible supply problem of depletion of natural resources and degradation of environmental services required for support of commodity production, and (ii) the resistance by communities and whole societies to the ecological and cultural depredations wreaked by expanding capital' (O'Connor 1994: 126). The introduction of markets and of market-simulating techniques is offered as the means of redressing the resulting production of externalities and is legitimated by reference to the growing scarcity of natural resources. Scarcity, indeed, offers another set of strategies for capital, through the exploitation of the produced scarcities created by the production of negative externalities. Witness the case of regions where surface and groundwater (and even tap water) sources are too polluted to drink (or are perceived to be so); worldwide sales of bottled water have grown rapidly over the past two decades.

Table 2.2. Nature as an accumulation strategy

Extensive accumulation	Intensive accumulation
Abundance	Scarcity
Nature	Environment
Resource exploitation (e.g. mining)	Nature capitalization (e.g. bio-prospecting)
Production (raw materials as inputs)	Consumption (leisure/recreation as outputs)
Growth - modernization	Growth – sustainable development
Material conquest of nature	Semiotic conquest of the environment

---

[3] The terms 'extensive' and 'intensive' employed here are distinct from those used in régulationist analyses. Although analogies can be drawn between the two uses of these terms, they differ in their referent and historical periodization. In the latter approach, 'extensive' accumulation has a spatial dimension, referring to the pre-Fordist strategies of mass exploitation, conjoined with the exploration, conquest, and consolidation of overseas sources of raw materials and outlets for produced goods. Similarly, the régulationist term 'intensive' refers to a (Fordist) regime of accumulation that is characterized by a deep, rather than broad set of strategies focused on the intensification of production, linked with real rising buying capacity, creating a 'virtuous' upward spiral of growth. As Altvater summarizes these concepts, 'the expansionist pressure inherent in the economic logic of surplus production has a territorial dimension (as production is necessarily always spatial). Surplus production is thus identical to economic conquest... of space, i.e. the 'production of space.' At first, space is conquered extensively; subsequently it is capitalized intensively' (Altvater 1989: 67). The analogy here would be that nature is first produced extensively, and subsequently capitalized intensively.

Scarcity is thus simultaneously deployed as a justification for privatization and commercialization, and exploited as an opportunity by capital. From a political ecological reading, proponents of privatization and commercialization are successful in blurring the distinction between 'second-order' (human-created) and 'first-order' or water scarcity, enabling the assertion not only of the desirability, but also the necessity and inevitability of privatization and commercialization. The threat of water scarcity is converted into an opportunity, through being deployed as a justification for the expansion of the market. To paraphrase Vandana Shiva, ecological modernization is 'capitalism's way of turning a threat into an opportunity' (Shiva 2002). Conservation should here be read in both a political economic and ecological sense: as the preservation of capitalism as a socio-economic system; and as the prioritization of environmental conservation, as market opportunity and strategic necessity, enacted by private companies via economic instruments. Water scarcity (as with other instances of the degradation of the conditions of production) is thus both structural to and highly functional for capitalism.

Externalities as 'territorial effects'

Neo-classical economics implicitly recognizes the contingent nature of scarcity in its definition as 'the property of being in excess demand at a zero price . . . [such that] a good that was never scarce would not be counted as an economic good at all' (Black 1997: 416). Without 'demand', i.e. human desires directed towards certain social ends, scarcity would not exist. What neoclassical theory does not explain, however, is how demand is constituted, why it changes over time, and why scarcity, as embodied in the scarcity postulate, should be a temporal and social universal, rather than a perception 'peculiar to the modern Anglo-European eye' (Xenos 1989: 2). From the neoclassical perspective, rent on natural resources is an index of natural scarcity that ensures, through variations in price in market exchange, the maximum efficiency of resource use. The price of commodities, as an index of natural scarcity, is a product of 'natural forces'. An environmental problem can in theory only exist whenever there is a gap between the '*accounting* price of a natural resources and its *actual* or *market* price' (Dasgupta 1990: 58 (original emphasis)). Such a gap, or 'externality', arises from imperfect or 'missing' markets, or from faulty government policy, or, more unconventionally, a failure to take temporal or intergenerational considerations into account (Helm and Pearce 1992). But why do such externalities so persistently recur, and most often with respect to so-called 'natural resources'?

Understanding why water is such an 'uncooperative' commodity, persistently characterized by externalities and hence market failure, requires reference to its biophysical characteristics; in other words, to its 'materiality'. Frequent reference is made to 'materiality' in political ecological debates, where the term serves as a kind of (rarely decoded) codeword for those seeking to (re)incorporate nature into political economic analysis. At one level, materiality refers to nature as an *object* of the analysis—an acknowledgement of the key role occupied by nature—transformed into resources—in our political economies. Water, for example, is an essential and to some degree non-substitutable factor of production in processes of agricultural and industrial production, and urbanization. This usage of the term 'materiality' implies an acknowledgement of the corporeality of our economies, of their embedding in natural processes.

The term 'materiality' also refers to an understanding of nature as a *subject* of political economic processes, whose specific biophysical characteristics shape the social relations of production, simultaneously enabling and constraining its own production. Conventional political economic analysis distinguishes between different kinds of 'raw materials,' the natural resources available 'free' to humans, only insofar as they figure in the production process (Benton 1989). As water flows through supply networks, for example, it is simultaneously a raw material (abstracted from a river), a product of the labour process (having been filtered, pumped, and chemically treated), and an instrument of labour (used not only in industrial manufacturing but also a physiological requirement of workers). This categorization is relational, determined by its specific function in the labour process. However, as Benton argues in his renovation of the concept of labour in an attempt to broaden historical materialism to address 'green' issues, analyses which rely on this categorization often overlook non-productive types of labour, thus evading consideration of the ecological implications of capitalist accumulation. In particular, the adaptive as well as transformative dimensions of the labour process are often overlooked in political economic analysis (Foster 2000).

Although I agree with Harvey's (1996) and Burkett's (1999) critiques of Benton's analysis, in particular the weakness in his resuscitation of a nature/society dualism, Benton's argument highlights an important point. The analytical focus of much political economy on labour as a transformative process rather than an adaptive process frequently allows political economists to ignore the very different qualities of various raw materials (Benton 1989), qualities that should be acknowledged in a relational-dialectic treatment of resources. Privileging the factory as archetypal worksite overlooks the different processes by which use-values are transformed into exchanges values in, for example, a forest or a river. Specific constraints imposed by different biophysical characteristics of 'natural resources' will give rise to specific issues in their appropriation into production which will affect how differential rents are captured. This is relevant to analyses which attempt to explain why some resources are more fully commodified, and seemingly more easily commodified, than others.

Water, for example, is a flow resource, not easily bounded above or below ground, through which externalities are easily diffused. It may serve multiple uses simultaneously, and be required to perform several functions in one circuit through the hydrological cycle; upstream users can greatly affect downstream uses. This difference is reflected in the fact that property rights are more difficult to establish for water, and boundaries often more blurred. To a first approximation, the monopolization of location (not water itself) in order to extract profit, and associated territorial effects apply to water as much as to land (Swyngedouw 1992). However, profit extracted from this monopolization of water, and to some extent its price, is determined through externalities (Swyngedouw's 'territorial effects') peculiar to it as a resource, in particular the degree to which negative externalities can be displaced through taking advantage of the unique flow properties of water. Because negative territorial effects are difficult to control or mitigate under private property regimes for flow resources, this provides a justification for the involvement of the public sector and eventually undermines private sector water provision.

Another biophysical characteristic of water that underlies its 'uncooperativeness' as a commodity is its density: water is one of the heaviest substances mobilized by human beings in their daily search for subsistence, and hence cheap to store, but expensive to transport relative to its value. Herein lies an important part of the reason why water historically has lain at the limit of the sphere of applicability of the market as a social institution for allocating resources. The public–private tension that besets water supply provision is due in part to water's biophysical characteristics: water is expensive to transport relative to value per unit volume, requiring large-scale capital investments in infrastructure networks which act as an effective barrier to market entry. The significant differences between water sources—and the negative ecological effects which may arise from mixing water from different sources—mitigate against inter-basin transfers, particularly those using existing surface-water channels as conduits. Water supply is thus a highly localized resource, and therefore highly susceptible to monopolistic control (economists' 'natural' monopoly). Fully commercializing water is as a consequence invariably fraught with difficulty. Yet, as the ecological modernists remind us, it is also ripe with opportunity.

Environmental externalities as an 'ecological fix'

Understanding why environmental degradation is (at different times and in different places) both useful for and a barrier to further capital accumulation requires reference to the crisis tendency inherent in capital accumulation. Disequilibrium is 'normal' to capitalism, which is inherently crisis-prone (Harvey 1985). Crises emerge from the contradiction inherent between the realm of production and the realm of exchange (Harvey 1982: 176). These crises take various forms at various scales, but may be expressed as crises of realization, resulting from a state of over-accumulation. This results in the devaluation of capital, which takes a variety of forms—excessive inventories, idle money capital, un(der)employed labour power, or unutilized productive capacity (Harvey 1982: 85; Harvey 1985). As Harvey outlines in detail, crisis can, however, be minimized or held off through a variety of strategies—capital switching between different sectors, for example (Harvey 1985). Temporal displacement of crisis through long-term investment, literally fixing capital, thereby slowing down its rate of turnover in order that the turnover time of the remainder is accelerated, is another strategy. These strategies, in certain cases, can result in what Harvey terms a (temporary) 'dynamic equilibrium', in which internal growth appears to be self-sustaining.

Critical to this appearance of equilibrium, and temporal displacement of crisis, is what Harvey terms a 'spatial fix'. Central to crisis-minimizing behaviour is 'socially necessary turnover time', the 'average time taken to turn over a given quantity of capital at the average rate of profit under normal conditions of production and circulation' (Harvey 1985: 135). The drive to command surplus labour time and convert it into profit necessarily implies the constant and ever-changing production of space by capital, for 'each form of the geographical mobility of capital requires fixed and secure spatial infrastructures if it is to function effectively' (Harvey 1985: 148). Location, in other words, is an 'active moment' within the overall circulation and accumulation of capital (Harvey 1982: 390). Harvey outlines how the contradictions of capitalism are necessarily expressed, at a variety of spatial scales (themselves produced in the course of capital accumulation), in fixed spatial structures that, originally produced to enable capital to overcome space, are themselves spatial barriers to be overcome (Harvey 1982: 430). Use values, in other words, have spatial properties; spatial relations (for example location) as well as the 'natural' properties of a commodity affect its value (Smith 1984).

Uneven geographical development—the production of global space as relative and differentiated space—and the linking of commodity production in different locations via processes of exchange, thus plays a vital role in crisis formation and resolution and thus in capital accumulation (Harvey 1982; Smith 1984). Capitalist development necessarily entails the 'continual transformation of natural space—inherited absolute space—into produced relative space' (Smith 1984: 87), and thus entails the production of space as well as the production of nature. Smith (1984) identifies an important parallel between these processes; just as the attempt to reduce spatial barriers and universalize the wage–labour relation levels human nature downwards, the capitalist pursuit of raw materials results in their degradation, manifested as declining quality of these materials (Smith 1984: 115). This 'equalization' process 'is manifested in the common scarcity of objects of labour' a scarcity, that is 'socially organised in order to permit the market to function' (Harvey 1973: 114). Scarcity, in other words, is socially produced as 'capital engages in a frantic search for the materials . . . which fuel the accumulation process' (Smith 1984: 115). Thus scarcity is both structural to and highly functional for capitalism; the scarcity of commodities is produced in a manner analogous to spatial relations such as location. This production of scarcity is an integral component of the production of nature, in which the quality of nature is levelled downward. So when Smith speaks of 'produced scarcity in nature' as the source of crises, he is speaking of socially produced crises, rather than crises emerging from a clash between society and an external, independent 'natural world' (Smith 1984: 60).

Drawing on O'Connor's and Benton's insights, we can postulate that the 'spatial fix' of capitalism has a parallel in an 'ecological fix', whereby capital extracts additional surplus value by degrading the conditions of production, thereby decreasing the socially necessary labour time entailed in the production of 'second nature' (O'Connor 1994; Benton 1989). In other words, the labour time embodied in any given commodity can be shortened through practices that degrade the quality of the conditions of production, and thus greater profit can be extracted. This is especially important in, but by no means exclusively applicable to what Benton (1989) terms 'eco-regulatory' and 'appropriative' labour processes, those in which labour 'proceeds as nature does herself . . . [and] in this work of modification [humans are] constantly helped by natural forces' (Marx 1990: 133–4). The degradation of the conditions of production—in the language of economics, the creation of (environmental) externalities—is thus highly functional to capitalism. Externalities, in other words, are only external to individual actors, but 'are essentially internal to the circulation of capital as a social process' (Swyngedouw 1992: 420). They are 'not rare or exceptional, [but rather] . . . an integral part of modern life, just as much as accidents and natural disasters are the other side of the Enlightenment coin of instrumental control' (O'Connor 1994: 142). Externalities are the inevitable by-product of our metabolism of nature, and under a corporate mode of provision they will be intensified in scale and scope.

This analysis, it should be noted, does not imply intentionality on the part of capital to degrade the environment, nor does it imply that the production of 'negative' externalities is always negative in economic terms. Externalities may both pose a threat to, and represent an opportunity for different capitals at different times in different places. The production of nature by capital thus creates conditions that simultaneously enable and constrain capital. The biophysical characteristics of an 'ecological fix' are thus neither simply 'good' nor 'bad' for capital; they both are mediated by and mediate capital accumulation insofar as they are produced as part of a landscape in which further social-economic activity occurs. This act of mediation is not only undertaken by private corporations; whether engaged in production or regulation, the state as well as private corporations may deliberately and actively resort to an 'ecological fix'. Both the public and private sectors, both the state and the market, can produce externalities; externalities are exacerbated by the corporate production of water (whether public or private). Nor does the production of externalities occur solely for economic reasons; the degree and distribution of externalities is socially contingent, related to technological development and cultural norms of water use, hygiene, and water science.

The dynamism of market environmentalism arises in its promise of simultaneously addressing and mobilizing water scarcity, in the pursuit of continued economic growth. It is important to note that transition to the 'market environmentalist' mode of regulation, as a response to the problems posed for corporate provision of resources under capitalism, is not enacted solely by the private sector. 'Capitalism' is here understood to include not only capitalists, but also the state. Market environmentalism is implemented by both private corporations and governments; corporations engaging in new strategic sectors and activities; the state reconfiguring its role as regulator (particularly environmental and economic), rather than provider, of environmental services.

Privatization as re-regulation

The notion that the state has abandoned the role of service provider and taken up a role of regulator arises frequently in discussions of the restructuring of utility services. What do we mean by regulation? The term implies a set of social relations or practices that shape and control economic activity—but precisely how do they do so? Economists generally define regulation as sets of rules together with procedures for their application and enforcement. Social scientists studying resource management often argue that formal rules are only a subset of the set of regulations (or institutions) governing resource management, and prefer to analyse *institutions*—understood as 'norms, rules, and customs' (Ostrom 1990). Political economists, in contrast, frequently employ the concept of a 'mode of regulation'—the dynamic (yet relatively coherent) set of social and political institutions that sustains a particular 'regime of accumulation'. Regulations and the act of regulating are embedded within a broader political economic process. Drawing on this latter distinction, from a political economic perspective we might define resource regulation as a practice of negotiating resource allocation, a practice both enabled and constrained by resource management institutions, by which people participate in the production and exploitation of resources.

If we define resource regulation as the social negotiation of resource allocation, terms such as 'deregulation' become a misnomer. Regulation does not refer merely to a quantity of rules or norms; it is, rather, a practice in which we always (and inescapably) engage. Redefining regulation in this way displaces the analyst's gaze. Rather than an analytical focus on a correct 'frontier' between state and market and a reliance on the public–private binary, attention is brought to bear on the diverse forms of regulation of which private corporate and public corporate (i.e. state) control are only two examples. Rather than perceiving privatization as an act of deregulation, attention is brought to bear on the ways in which the state engages in institutional learning and accordingly enacts change in regulatory frameworks; the state does not necessarily withdraw, but rather changes the nature of its interaction with citizens and corporations.

In some cases, these changes may entail deregulation (a reduction in the number and scope of rules and regulations applied to commercial activity); in other instances, the oversight role of the state may intensify. As Pierson notes, 'while the state may increasingly act through *regulating* rather than actually *delivering* services, at the same time it may actually become more active and intervene more intensively (and intrusively)' in the activities of corporations and the lives of citizens (Pierson 1998: 165). Privatization, for example, may be accompanied by an increase in regulatory surveillance and information demands (as has indeed been the case in the UK). This may occur despite the intentions of the rule-makers; particularly, as the latter half of this book explores, for resources which are difficult to commercialize.

Two sets of ideas about the state are implicit in this argument. First, the welfare state is characterized by competing and to some degree incommensurable claims from capital, on the one hand, and from citizens, on the other; a failure to balance these claims may be perceived as a crisis (fiscal, or ecological, and/or political). Offe characterizes these claims as 'settlements'. The politico-economic settlement establishes a relationship between the state and economy generally supportive of capital accumulation, but incurs costs which must be mitigated by the state, requiring a second, politico-normative settlement establishing a relationship between the state and civil society, in an attempt to ensure political legitimacy and stability (Offe 1984; 1985). These two settlements are mirrored by a functional separation of tasks within the welfare state, divided into economic, politico-administrative (administrative and institutional structures), and politico-normative (organizations providing state welfare) subsystems, that offer divergent and sometimes mutually contradictory goals, conceptions of social justice, and modes of calculation. The state makes decisions in the collective interest of continued economic growth, and is rewarded with fiscal inputs (via taxation) that sustain its existence and fund regulatory functions (Hay 1996). However, in order to secure the social legitimization necessary to be able to carry out regulation, the state provides welfare services, mitigating the inequalities of market outcomes (Hay 1995). Regulation succeeds when it negotiates a compromise between these competing settlements; failure (or perceptions of failure) is one of the key contributing factors in macroeconomic and political crisis in liberal democracies (Haughton 1998; Hay 1995).

Within the contested process of re-regulation which ensues, the state, as an ensemble of forms of government and institutions of governance, is a strategic terrain, or what Jessop terms a 'site of struggle': an object and generator of strategies as well as the product of past political struggles (Jessop 1990). As a 'key site in the strategic codification of power relations' (Jessop 1990: 248), the state plays an active role: generating political strategies in mediating the process of re-regulation; and articulating sectors and scales of economic activity. Privatization, for example, is in many instances a strategy undertaken by the state. Post-privatization, the competing claims of capital and citizens continue to be mediated by the state, formalized as (often incommensurable) demands concerning the structure and functions of the regulatory framework. Greater attention to the fine-grained analysis of processes of socio-economic restructuring, predicated on an understanding of regulation as social practice as well as economic imperative (echoing Clark's approach to 'real' regulation (Clark, 1992*a*; 1992*b*), clarifies the links between a macroeconomic regulatory framework and the practice of rule-making, and thereby between different scales of economic activity. The socio-cultural characteristics of British regulation—traditionally non-legalistic, informal, and discretionary (Foster 1992)—are, for example, as important as the stated agendas of governments and private water companies in understanding the original structure and post-privatization evolution of the regulatory framework for network utilities.

As subsequent chapters explore, the evolution of resource regulatory frameworks entails interrelated processes of socio-economic restructuring and environmental change. This re-regulation is simultaneously political, economic, and material—with very real implications for the environmental conditions of existence for humans and non-humans alike. Yet resource re-regulation is, at the same time, representational: an act of interpretation as well as mediation, involving both categorization and adjudication, and requiring problem definition as well as problem solving. Following Castree and Braun, we can draw on Heidegger's concept of 'enframing' to argue that nature and our experience of it are 'related to a configuration of historically specific social and representational practices which form the nuts and bolts of our interactions with . . . the world' (Castree and Braun 1998: 17). Resource regulation is inherently (but by no means solely) a discursive practice, as well as an institutional framework embodying rules that define knowledge and legitimize authority. Here, discourse is defined as 'a specific ensemble of ideas, concepts, and categorizations that is produced, reproduced, and transformed in a particular set of practices and through which meaning is given to physical and social realities' (Hajer 1995: 44). Discourse is constituted by practices through which concepts and categorizations, as well as narratives, are (re)produced and transformed (Hajer 1995). Discourse does not exist in isolation; it is embedded within particular institutional configurations of power, knowledge, and accepted authority—in short, within competing ideologies. Where ideology serves a legitimating function by offering 'sets of reasons for material conditions' (Eagleton 1991: 209), discourse is a practice by which ideologies are created, deployed, and contested.

Defining regulation as the social negotiation of the exploitation of a dynamic resource landscape thus requires an analysis of the relationship between the discursive, social, and material dimensions of environmental conflict and change. It is in this sense of 'deep discursivity', as well as a social and material sense, that regulation entails the production as well as exploitation of resources. As such, regulation is enacted by individuals, or groups of individuals, who share common 'storylines'—sets of (often contested) ideas which unite people in particular ways not only of communicating about, but also of producing knowledge of, a problem, issue, or event. Regulation is thus inevitably inscribed within ideological allegiances, as well as political alliances.

Here we can begin to characterize the constantly evolving character of regulation: as a practice, regulation is 'inscribed from within with a multiplicity of ideological 'accents'; and it is in this way that it sustains its dynamism and volatility' (Eagleton 1991: 195). This dynamism is inherent, rather than incidental, as regulation is one of the principal stages upon which (at least in democratic, capitalist societies governed by well-developed bureaucracies) social negotiation of contradictions inherent in the accumulation–regulation nexus is played out. Approaching privatization as a process of ongoing re-regulation enables an exploration of the creative, seemingly chaotic transition from one mode of social regulation to the next (Hay 1995). Accordingly, Chapter 3 now turns to an analysis of the transition between the 'state hydraulic' and 'market environmentalist' paradigms of water management.

Reflections

Extensive political economic analysis has been undertaken of: the privatization of public services; the restructuring of state functions, and the 'greening' of economic and political life in the final decades of the twentieth century. This chapter has brought some of these debates to bear on the ongoing transition in practices of water management. In doing so, it has attempted to demonstrate the shortcomings of, and move beyond both the political economic and neo-liberal depictions of the 'crisis' (and ensuing 'retreat') of the welfare state. It has also attempted to move beyond the 'thick description' of many ecological modernization perspectives, by presenting a conceptual framework for analysing the relationship between socio-economic restructuring and environmental change.

On the one hand, where water resources and supplies mobilized under the 'state hydraulic' paradigm, they both contributed to and were affected by fiscal crises at the macroeconomic level; in attempting to stave off these crises, states may resort to an 'ecological fix', underinvesting in water infrastructure and permitting the multiplication of environmental externalities. Market environmentalism arises in part as a response to this exhaustion of the 'ecological fix', as states strategically internalize externalities (eg. via commercialization), and/or seek to divest fiscal responsibility for the water sector (e.g. via privatization). It also arises as a process of institutional learning (or 'reflexive self-confrontation', as the risk society theorists would have it), in which both state and capital variously seek to limit the growth of, yet also seek new sources of profit in the reduction or containment of environmental externalities. Market environmentalism can be thus rephrased: water is dually subject to processes of the intensification of nature accumulation and the 'marketization' of the public sphere, enacted through a process of re-regulation which is simultaneously material, socio-economic, and discursive.

Privatization, then, is part of a much broader (and deeper) process of re-regulation of our relationship with nature, and one another. It is an adaptive as well as transformative process, which must contend with water as an 'uncooperative' commodity—whose biophysical characteristics render its commercialization difficult, at least when using the technology of networked water supply provision which is the norm in the industrialized 'North'. The implications for government, regulators, and the water supply industry of this 'uncooperativeness' are considered in subsequent chapters, and in particular in Chapter 7.

This chapter has also attempted to outline what a political ecological analysis might contribute to political economic studies of water supply privatization. Specifically, it has discussed three ways in which political ecologists seek to recover certain aspects of the political economy tradition in examining the mutually constitutive relationship between socio-economic restructuring and environmental change: acknowledging the materiality of nature; re-theorizing resource regulation; and interrogating the role of the state from a different perspective than that of much political economy.

My claim here is not that the 'interrogation of the state' is a move unique to political ecology. Rather, political ecologists approach the state in a manner somewhat distinct from political economists. More precisely: the 'retreat of the state' is a very ambivalent process when its environmental impacts (rather than the redistribution of the social surplus amongst humans) are considered. The state has in some cases rationally administered massive environmental degradation and systematic under-provision of environmental goods. In the case of water, some of the great gains in human welfare during the twentieth century associated with the 'state hydraulic' mode of regulation were made at the expense of the environment—with the state temporarily devolving costs on to the environment (the 'ecological fix'). Privatization and commercialization of water often occur together with a simultaneous commodification and (re)valorization of the environment—prioritizing environmental protection over other uses (such as industrial uses) or priorities (such as affordability).

The 'market environmentalist' paradigm, when applied to water management, produces clear gains for the environment in some cases; hence the frequent disagreements between environmental groups and consumers groups in contemporary debates over water privatization. Subsequent chapters will concern themselves with this expanded definition of the 'subjects' of distributive justice, and examine the implications of the privatization and commercialization of water supply for consumers, the environment, and the water industry. The role played by the state in excluding the market from water provision in England and Wales, and then subsequently initiating and enacting market environmentalism, will be examined in Chapters 3 and 4, which detail the consolidation of the 'state hydraulic' mode of regulation and the gradual emergence of 'market environmentalism' in water supply management. The implications of this 'great transformation' for consumers, the environment, and the structure of the water supply industry are considered in Chapters 5, 6 and 7 respectively.

3

Building the Networks

For many people around the world, water is an artisanal product, drawn from local streams, wells, or ponds, or delivered by water vendors in jerrycans or tanker trucks to the home. A farmer in Cambodia, or a slum-dweller in Jakarta, might use 25 litres per day. In industrialized societies, consumption levels range from six to twelve times this amount; water supply is an industrial product; produced, distributed, and consumed in large quantities, to generic quality standards. Large amounts of water—a substance we consume daily in greater volumes (indirectly as well as directly) than most other goods—are taken for granted as a necessary requirement for daily life in modern, industrialized societies.

The emergence of an industry dedicated to the mass production of water, and the concurrent conversion of water from an artisanal to an industrial product, occurred in Britain during the nineteenth and early twentieth centuries, a period of unprecedented urbanization and industrialization. It was during this period that the water industry achieved near-total control over water abstractions, that water rights were de-linked from land rights, and that the majority of consumers came to rely on treated water supplied through vast networks, rather than untreated local sources. Private companies were active in the water supply industry throughout this period, and dominated the industry in the mid-nineteenth century. By the mid-twentieth century, however, the water supply industry was almost completely publicly controlled, and remained so until the privatization of the English and Welsh water supply companies in 1989. The Victorian experience with private water supply is critical to understanding the subsequent growth of the public water industry. Accordingly, this chapter returns to the nineteenth century, and examines the growth and decline of private water companies before documenting the emergence of municipal water supply systems and their progressive rationalization and centralization, culminating in the nationalization of water supply production in 1974. The gradual commercialization of the nationalized water supply companies is detailed, with emphasis on the fact that privatization was only implemented after a sustained programme of commercialization of water supply services throughout the 1980s.

Much of the analysis in this chapter focuses on the building of water networks. Water infrastructure networks have one of the longest turnover times of any dedicated utility infrastructure; over one-third of the pipes supplying water to London, for example, are more than one hundred years old (Perera, Farley *et al.* 1985). Given this longevity, water networks, as they age, come to materially embody successive phases of capital accumulation. The fixed capital embodied in the material network of mains, pipes, pumps, reservoirs, and sewers is an artefact of the hydrosocial cycle, shaped by successive generations of social, technical, and economic practices. For example, although most private water companies had been absorbed into municipal water trading services by the mid-twentieth century, some private companies—usually in urban areas—were allowed to continue operating. This is one of the reasons for the incomplete spatial coverage of the later regional water monopolies, and the fragmentation of supply networks in the greater London area, still served by five different companies. Contemporary water networks reflect historical choices and practices. There is thus nothing 'natural' about how and where contemporary water networks in England and Wales are found. Today, water infrastructure networks are socio-economic monopolies that have become 'natural', or rather naturalized. In this chapter we follow the building of those networks as part of the process of the urbanization and municipalization of water production.

Making connections: urbanization and industrialization of water supply

We are in an era of giganticism, where nation-wide economic survival depends on gigantic organizations for mass production. We no longer irrigate our own smallholdings and water our cattle from the village pump; we mass together for work, we mass-produce in mammoth factories . . . and therefore, we must mass produce our water. (Twort 1963: 9)

The imposition of regulatory control over water supply must be understood in the context of rapid urbanization and associated industrialization of water supply. In the nineteenth century, few people anywhere in the world had running water through taps in their homes. Wealthy urban residents had access to networked provision supplied by private companies; most other urban residents fetched water from common taps (standpipes) or purchased water from street vendors. Urban residents' methods for supplying water and disposing of sewage were inherited from the medieval town, whose methods of water supply and disposal of wastes were best-suited to a relatively limited, spatially contained urban population. These methods were ill-suited to the large and rapidly growing cities of nineteenth-century England. In the medieval city, wastes or 'nightsoil' were often employed as fertilizer, and were viewed as part of the natural metabolism of the city. But with the advent of large-scale urbanization, the smell of human excrement began to lose its rural associations with fertility, and was increasingly viewed as indicative of disorder and decay.

According to the theory of 'emanations' or miasmic theory of disease widely held at the time, foul odours were carriers of disease; hence the imperative for the removal of the sources of odours from cities. The rationalization of urban space that occurred in many cities in the nineteenth century and which gave rise to the grid-like city form that we are familiar with today was in part a reordering of the flow of wastes through the urban environment.

The development of waterborne sewerage systems was also supported by the increasingly influential scientific conceptions of the germ theory of disease, linking water-based hygiene with diseases such as cholera and typhoid. Campaigners such as John Snow in London sought to change public hygiene norms, largely sparked by the profusion of water pollution due to lack of sewage disposal in urban areas, and the resulting outbreaks of cholera and typhoid which were common in urban areas in the eighteenth and nineteenth centuries. At the same time, views towards personal bathing were changing; bathing, which in Europe in the Counter-Reformation period had been associated with pagan spirituality or degenerate court behaviour or poor morality, became instead associated with physical and moral health. Washing became an important activity, but was no longer collective, as in the Roman period, but an individual activity, in a private space—and separate space was created for washing within the home—the bathroom is a relatively recent invention. This was central to the emergence of differentiated domestic, but also new private spaces—such as the bathroom—that characterize the modern home. Water was not only enrolled as a means of flushing the city—a vehicle for cleaning and ordering the city—but also played a key role in the 'sensory realignment of modernity' (Gandy 1999).

Urbanization was thus a cultural, as well as material condition catalyst for the building of the large-scale water networks, as well as for the resulting industrialization of water supply. With the transition from artisan to industrial means of water production, community control was gradually ceded to corporate control. Corporate control of water supply, in other words, was enabled by the creation of economies of scale and population densities sufficiently high to enable large-scale water production to be economically viable. That the most efficient scale of operation was a single provider—or natural monopolist—tended to encourage the emergence of water corporations. The difficulty of enforcing long-term contracts—typical of many services but particularly acute given water's unique biophysical characteristics, which necessitated large up-front capital investments and long life-span of infrastructure—was an additional incentive for the emergence of corporate provision of water supply. In the context of Britain's nineteenth-century *laissez-faire* approach to the economy, the corporate providers which emerged were largely private.

From private to government control

Why was corporate regulation of water resources increasingly concentrated in the hands of the state, rather than private firms? One reason was that water infrastructure played a symbolic role in the territorialization of state power at the national level, and in the reinforcement of civic identity at the municipal level. Dam-building and the harnessing of large-scale water resources was of critical importance to national elites; the creation of water supply infrastructure often played an analogous role in the development of municipal governance. More importantly, perhaps, the political consensus that had emerged by the late nineteenth century on the need for universal provision (as outlined above) rendered provision of water supplies less profitable for the private sector, which typically concentrated on wealthier consumers. Calls for state intervention to ensure universal provision were based in part on the recognition of the under-consumption of water and sewerage supplies—and the negative public health impacts— that resulted when provision was left to private companies and individual consumers. Water supply and sanitation, argued proponents of state control such as the 'gas and water socialists', conferred greater benefits on society than captured in people's own preferences for them; these 'merit goods' would be under-consumed if left to individuals and the market. According to this argument, water and particularly sanitation, like other merit goods such as education, should be subsidized or provided by the state.

The under-provision of merit goods in nineteenth century England posed high risks in terms of public health (particularly sporadic cholera and typhoid epidemics) in rapidly growing cities. The societal response, across a number of utility and service sectors, was to organize provision collectively. However, the universal provision of water and sewerage posed difficulties due to the capital-intensiveness of the networks chosen as the preferred technology of provision, which were characterized by low levels of profitability when universal provision criteria were imposed. Water and sewerage might be categorized as what Offer terms 'prudential goods', requiring large sunk costs, with staged delivery over a long period of time implying benefits were not immediate (or 'visceral') but rather durable (Offer 2002). If the high public health risks of under-provision were a justification for state involvement through the collective regulation of provision (merit good), the requirement of a large initial investment over time combined with the low profitability implied when universal provision requirements were imposed favoured state rather than private ownership (prudential good).

At first, as Offer (2002) notes: 'in Britain, prudential goods were originally provided by private investors for profit, and by non-profit voluntary "clubs" of exclusive constituencies' (Offer 2002: 4). Yet beginning in the last quarter of the nineteenth century, the state became increasingly involved in the provision of prudential goods, to the extent that 'between 1870 and 1970... collective provision grew faster than the economy' (Offer 2002: 4). Private provision was thus gradually displaced by public provision, either through public ownership or highly interventionist state regulation. In Britain, municipal governments gradually took over the ownership and responsibility for management of water systems, as with other utility services. The few private companies that remained were tightly regulated. A similar pattern was followed in the United States and Canada. In France and Spain, asset ownership was taken over by local governments, but municipalities in some cases chose to contract out network management to private water services companies. Across the industrializing world, with few exceptions, asset ownership of water supply and sanitation networks gradually became the preserve of the state. In Britain, however, the state hydraulic paradigm began to break down by the late 1970s. To understand why, we return to the nineteenth century, at the time of the building of the networks.

Laissez-faire: private provision for the public good (*c.* 1750—1870)

Ninety-nine per cent of the residents of England and Wales are connected to the public water supply system, their supply being piped directly into their homes from sources that may lie hundreds of kilometres away. This is one of the highest connection rates in the world (Table 3.1). The degree of market concentration is also one of the highest in the world.[1] In England and Wales, water for domestic consumption is an industrial product, abstracted, purified, and shipped to the consumer by licensed water companies in England and Wales which collectively operate nearly 300,000 kilometres of pipeline dedicated to water supply alone (Thomasson 1993). Sewerage services are similarly networked; England and Wales have one of the highest percentages of the population connected to the sewer system in Europe (WSA 1992).

The industrial production of water, and its supply to the vast majority of the population of England and Wales, is a relatively recent development. In the mid-nineteenth century, less than 10% of households had a piped water connection (Gregg 1950). Yet by the 1950s, over 90% of the population had access to piped water (Sleeman 1953), supplied by networks bringing vast amounts of water over long distances; from the Pennines to east Yorkshire, from the Cotswolds to London, from Wales to Manchester and Liverpool.

---

[1] In Germany and the United States, for example, the water industry is highly fragmented, with thousands of small, local providers. In France, the water supply market is dominated by a few very large multinational firms; unlike English firms, French firms do not own the water supply infrastructure. Municipalities retain ownership of the assets, and award long-term 'affermage' or concession contracts to the private firms. In countries with a lower degree of urbanization and industrialization, a complex mix of public and private ownership and supply, in which industrial and artisan modes of water production co-exist, is the norm.

Table 3.1. Area and resident connected population, water supply and sewerage services (1998/9)

	Water services		Sewerage services	
Water service companies[1]	Area[2] (km^2)	Resident connected population (000)	Area (km^2)	Resident connected population (000)
Anglian	22,000	4,044	27,500	5,418
Dwr Cymru	20,400	2,855	21,300	2,954
Northumbrian	8,993	2,534	9,400	2,630
North West	14,415	6,854	14,445	6,799
Severn Trent	19,745	7,331	21,650	8,225
Southern	4,450	2,219	10,450	3,991
South West	10,300	1,509	10,800	1,396
Thames	8,019	7,634	13,302	12,423
Wessex	7,350	1,172	10,000	2,420
Yorkshire	13,900	4,521	13,600	4,737
Subtotal	129,572	40,673	152,447	50,993
Water supply companies[3]				
Bournemouth and West Hampshire	1,041	430		
Bristol	2,391	1,050		
Cambridge	1,173	288		
Chester	142	108		
Essex & Suffolk	2,850	1,681		
Fokestone and Dover	420	158		
Hartlepool	90	92		
Mid Kent	2,050	555		
North Surrey	500	478		
Portsmouth	868	650		
South East	3,607	1,377		
South Staffordshire	1,507	1,233		
Sutton & East Surrey	834	633		
Tendring Hundred	352	144		
Three Valleys	3,200	2,432		
York	340	175		
Subtotal	22,055	11,632		
Total	151,627	52,305	153,516	50,993

[1] Licensed for both water supply and sewerage services
[2] Area includes inland waters
[3] Licensed for water supply services
Source: WSA 2000

The oldest sections of these networks had, in many cases, been built by the private sector. Many of the first water suppliers in Britain were private joint-stock companies (Millward 1989). It was a private company—the New River Company—that built the canal that permitted the first long-distance transfer of water into London, over a distance of 40 miles, in the early seventeenth century (Metropolitan Water Board 1949). By the end of the seventeenth century,

London had 20 water conduits, mostly built by individuals or private companies; many were supplied by the London Bridge pumping station, built in 1582, after which 'the principle of granting a water supply to private properties was adopted' (Goubert 1986: 172).

Private water supply companies were set up throughout the seventeenth and eighteenth centuries, tending to operate in urban areas, where economies of scale were more easily achievable. In London alone, there were nine private water supply companies by 1828 (Goubert 1986: 172). These companies filled in, or covered over open water networks, which had for centuries supported a multiplicity of uses (for example leisure, transport, trade, drinking water, and effluent disposal). This burial of water supply networks was simultaneously an enclosure of the water 'commons', which established a series of (occasionally overlapping) monopolies, largely concentrated in urban areas, with diverse technologies, water sources, and consumption patterns (Dimcock 1933). Water supply outpaced sewerage, the latter remaining less financially attractive and less widely distributed, in part because of the established practice of spreading human wastes on land rather than disposing of them through water—indeed, until 1818, the law prevented waste other than from kitchen and surface sources from entering sewers (Newson 1992).

Municipal governments also had a long tradition of involvement in public water supply, but they had largely given up these powers by the late eighteenth century, as the delivery of water was increasingly regarded as a normal commercial venture (Hassan 1985). Parliament discouraged municipal or community involvement in water supply (for example in Manchester in the early nineteenth century) and frequently reiterated its preference for private commercial ventures (Hassan 1985), encouraging competition in the belief that this would increase efficiency (Sleeman 1953). Another factor in the lack of municipal involvement in water supply was the hapazard legislative framework in the early phases of development of the water industry, in common with other utility industries. There was no national water legislation until the Waterworks Act of 1847. As each water supply initiative was created by an Act of Parliament initiated by a Private Member's Bill (Foster 1992), and as few municipalities had the resources to lobby for such a bill, private suppliers dominated in the first half of the nineteenth century. Some water supply systems remained in public hands; in some towns (such as Bath) the rights to ancient springs were owned and operated by the municipal corporation, or leased to private individuals, but by 1851, private water companies covered over 60% of the urban population which had access to piped water supply (Table 3.2).

This complex institutional mix was in part due to the institutional framework of riparian rights established under English common law, which recognized the landowner's right to use surface water running through his or her land, but allowed sale only when the water was not flowing in defined channels (Getzler 1993). This interrelationship between land rights and water rights ensured continued conflict, as access to new water sources required gaining

Table 3.2. Water supplies in 81 leading provincial towns and cities in Britain, 1801–1901

Year	Without Parliamentary authorization, or unknown			Municipal works established by Act of Parliament			Joint stock companies established by Act of Parliament		
	Works Number	Towns % total	% population	Works Number	Towns % total	% population	Works Number	Towns % total	% population
1801	69	85.2	79.9	7	6.2	10.3	5	8.6	9.8
1811	66	81.5	68.0	7	9.9	10.2	8	8.6	21.8
1821	62	76.5	60.3	8	13.6	10.4	11	9.9	29.3
1831	50	61.7	43.9	10	26.0	11.3	21	12.3	44.8
1841	40	49.4	33.5	10	38.3	10.7	31	12.3	55.8
1851	21	26.0	15.4	16	54.4	23.4	44	19.6	61.2
1861	7	8.6	2.9	33	50.6	46.8	41	40.8	50.3
1871	3	3.7	1.1	45	40.7	60.1	33	55.6	38.8
1881	-	-	-	65	19.8	79.9	16	80.2	20.1
1891	-	-	-	70	13.6	85.1	11	86.4	14.9
1901	-	-	-	73	9.9	86.9	8	90.1	13.1

Source: Hassan (1985)

access to land, and as an increase in water-intensive industries such as textile mills affected downstream water quality and quantity. Waterpower played a critical role in the power supply of the Industrial Revolution, more so at first than steam (von Tunzelmann 1978); control of water sources was thus highly contested (Hassan and Taylor 1996). By the mid-nineteenth century, in the textile districts of the North and dense industrial conurbations emerging in various parts of Great Britain, local water sources were frequently unfit for drinking purposes, and even unfit for industrial use (Smith 1972; Hassan and Taylor 1996). As municipalities and industry sought water sources further afield, conflicts amongst water users intensified (Hassan and Taylor 1996), and a 'scramble' for control of upland gathering grounds ensued (Smith 1972: 27). Mill owners and other industrialists represented a powerful lobbying force, and were successful in ensuring claims for the guaranteed release of water to 'compensate' them for disruption of flow levels caused by impounding upstream (Risbridger 1963).[2] Indeed, the interrelationship between land and water was to remain a source of conflict between competing users into the twentieth century; the link between water rights and property rights was finally severed only in the 1960s with the legislative requirement to obtain licences for all water abstractions.[3]

Favouring a *laissez-faire* approach and thus reluctant to regulate water provision, Parliament nonetheless found itself drawn into a regulatory role as

---

[2] Notification was required to be given to all industrialists using a river for a distance of 32 km below every proposed water supply scheme, who lobbied successfully for 'compensation' flows (as they are still known), which in some cases varied to coincide with working hours at the mills (Smith 1972).

[3] According to the provisions of the 1963 Water Act small-scale abstractions (currently below 20 m^3/day) are exempted from the licence requirement.

competition between domestic water supply and industrial uses persisted throughout the latter half of the nineteenth century.[4] The Waterworks Clauses Act of 1847 was the first general Act standardizing waterworks practice throughout the country; previously, individuals or local bodies had supported the passage of private bills through Parliament with respect to specific water supply ventures. The 1847 Act was the first to lay down general principles of, and guidelines for water management, water supply, mines, water rates, water pollution, sewerage (discharge of cesspool waste into urban drains became obligatory), and construction of waterworks. Together with its 1852 and 1863 amendments, the 1847 Act constituted the core of water supply legislation for over a century. In practice, regulation was limited to fairly superficial matters, providing no framework for the protection of wider environmental or consumer interests (Hassan and Taylor 1996). Provision was made, however, for regulation of maximum charges and dividends of the private water companies, for the protection of the rights of property owners, and for the protection of public safety (Sleeman 1953).[5] The Acts thus signalled the first formal recognition of the failure of the English experiment with *laissez-faire* in the water industry (Hassan 1985); notably, with respect to water supply, a precedent for universal provision was established. The statutory water undertaker was mandated 'to provide and keep in the pipes . . . a supply of pure and wholesome water sufficient for the domestic use of all inhabitants' (cited in Pugh 1939: 287), provided recipients were able to pay, and provided no private service existed to an adequate standard.[6]

This goal, however, was not soon attained. In the expanding cities of the nineteenth century, in which water's evolution into an industrial product began, supply networks were sporadic, and subscription was selective rather than universal (Gregg 1950). In the 1840s, perhaps 10% of the population had a piped water supply.[7] For the poor, water was still an artisanal product, purchased from private vendors in a few large cities (Dracup 1973*a*; 1973*b*), or collected individually from wells, rivers, and public pumps (for which they often also paid) (Chadwick 1842).[8] The very poor begged or stole water (Gregg 1950). Adequate sanitation was almost completely absent, with well-documented public health implications.[9] Municipalization of water supply services in the second half of the nineteenth century stemmed in part from concern over these public health issues, as explored in the following section.

---

[4] A survey by the County Councils Association in 1899 found conflict between local consumers and colliery owners or water companies in almost every county in the country (Sheail 1982).

[5] Some of these categories of regulation were only fully developed in the Gas and Water Facilities Act (1870).

[6] Cmd. 5986 *Second Report of the Central Advisory Water Committee: Consolidation and Amendment of the Law Relating to Public Water Supply* (1939).

[7] Cmd. 6513. *A National Water Policy* (1944).

[8] In London, for example, water was available for purchase by the bucketful (Chadwick 1842).

[9] See, for example, Chadwick (1842); Snow (1936, orig. 1855).

Municipalization (*c.* 1870–1945)

Amidst the rapid urbanization of the Industrial Revolution, the limited availability of clean water supplies led to widespread concerns over water quality and availability (Buer 1926). Investigations into the numerous cholera and typhoid epidemics of the nineteenth century established links between polluted water and public health (for contemporary reports, see Chadwick 1842; Snow 1936, orig. 1855). Strong links were made by an alliance of the medical establishment and social reformers between clean material surroundings and moral rectitude (Luckin 1986). These reformers linked physical and moral 'cleanliness', espousing a vision of a moral order in which 'physical wellbeing, the preservation of the family and the maintenance of morality' were essential to the 'security of society' (Goubert 1986: 229). Water supply was recognized not only as 'a public utility that was important not only for reasons of health and hygiene, but also for economic, agricultural and industrial reasons' (Goubert 1986: 185). Clean water was recognized to be a critical element for continued industrial production and a functioning labour force; universal provision was thus an economic, as well as social imperative.

Private sector provision, however, was not regarded as a likely vehicle with which to achieve universal provision. The standards of service being formulated by public health campaigners—universal provision of clean water twenty-four hours per day—would not be met without a significant increase in regulation to control not only standards but also prices and dividends. Political perceptions of market failure, and the pitfalls of private monopoly supply of basic goods, buttressed by experience of regulatory failure, lent further support to moves to secure public ownership (Hannah 1992). Advocates from a variety of backgrounds, from the 'gas and water socialism' of the small but influential Fabian Society, to utopian socialists and philanthropists supported public ownership of water supply (Fraser 1973; Goubert 1986). Collectively, these factors provided a strong justification for the municipalization of water services (Hassan 1985).

Equally compelling was the fact that private water supply was not necessarily a highly profitable enterprise. Competition amongst privately owned water companies in the larger industrial towns lowered profit margins, particularly in areas with direct competition, where two sets of mains might be found running down the same street (Sleeman 1953). The typical result was an increase in cost and decrease in efficiency when two expensive networks, neither used to capacity, were installed. Faced with a falling rate of profit, private water companies often 'squeezed' their customers where possible (for example charging more for delivery to upper floors), and 'cherry-picked' the most profitable customers. Soon, however, many water companies had fully penetrated those segments of the market that could afford their services, and expansion of water networks slowed. Many companies were barely profitable, and some did not even declare a dividend (Millward 1989). Increasing average costs due to pollution or exhaustion of local water resources was another critical factor in the decline in the profitability, and in the number of private water companies. Although the statutory companies' financial performance was regulated to some degree (for example with a ceiling on the maximum price per unit volume they could charge, no legislative framework existed for 'externalities' such as water quality.

With the introduction of new legislation permitting municipal entry into water 'trading', the dominance of private companies in water supply began to decline, having reached its peak in the mid-nineteenth century (Table 3.3). A handful of water companies figured in the top 100 British companies by ordinary share-market capitalization in 1889; by 1913, the list contained no water companies (Kennedy and Delargy 2000). At the same time, the number of municipal water undertakings increased. Although fewer than 25 towns per decade adopted municipal supplies between 1846 and 1865, over 60 per decade did so between 1865 and 1905. By 1907, 81% of the output of waterworks was controlled by local authorities (Sheail 1986: 53). The number of municipal water undertakings increased quickly: in the 1830s, there were 11 municipal water undertakings in Great Britain, and by 1914 over 700 (Millward 1989).

Table 3.3. Reduction in number of water services suppliers, 1945–1966

Year	1945	1956	1958	1959	1960	1961	1962	1963	1964	1965	1966
Number of undertakings	1,174	1,025	993	913	781	720	601	488	399	363	318

Source: Prusmann (1968)

The rapid growth in the numbers of municipal water undertakings was facilitated by the Gas and Water Facilities Act (1870). The 1870 Act enabled municipal water 'trading' companies to be created through a simpler process of administrative decision, rather than the laborious (and advantageous to private capital) process of Parliamentary Bills (Newbery 1999). The incentive for municipalities to engage in water trading was provided by the Public Health Act (1875), which made local authorities responsible for ensuring adequate and satisfactory water supplies. Municipalities were thus encouraged, and endowed with legal authority, to take over private water companies on public health grounds. The opportunity for municipalities to do so was facilitated by Parliament's practice, common after 1870, of granting limited life franchises (typically 21 years) to any privately owned network utility, at the end of which the municipality had the right to purchase the private utility (Newbery 1999).

Municipal involvement in water supply, as intended by the enabling legislation, complemented and then gradually displaced the activities of private water supply companies. In 1910, there remained 200 statutory joint-stock private companies, and 84 other private water companies, contrasted with 786 local authorities and 35 joint water boards operating water supply systems (apart from 1000 private proprietors generally supplying their own needs) (Dracup 1973*d*; Millward 1989). The predominance of municipal governments in water provision was higher than for gas, trams, or electricity (Newbery 1999). The private companies were spatially concentrated in the South, and particularly in and around London, reflecting their emergence in urban areas, but also the relatively greater impact of municipalization as a political movement in northern England (Sleeman 1953). In terms of domestic consumers, a major shift had taken place; by the early twentieth century, approximately 80% of water consumed was supplied by local authorities (Dracup 1973*c*; Pugh 1939). This was reflected in local authority debt figures; in the latter part of the nineteenth century, expenditure on waterworks represented one third of total local authority debt (Ashworth 1960: 226–7). This high debt load was incurred in part because 'supply of water was regarded almost as a civic duty on the part of local government. Thus water, by most criteria, was one of the cheapest commodities available to the population, being collected, treated and distributed to consumers for only a few pence per ton' (Hassan 1998: 61) It was also established practice for local government to subsidize sewerage through the rate-support grant; no provision was typically made for renewal of ageing assets, a policy that was to have important implications for the sustainability of the industry in the post-war period (Parker and Penning-Rowsell 1980).

Municipalization was part of a part of a broader devolution to, and expansion of muncipal powers in the nineteenth century (Hassan and Taylor 1996; Millward 1989), and was further encouraged by the emerging perception that public utilities differed substantively from other industries, in that a substantial degree of monopoly (rather than encouragement of competition) was required for efficient service (Sleeman 1953). The promotion of competing undertakings 'was abandoned, and each undertaking was given a monopoly in its own area' (Sleeman 1953: 33). With the goal of universal service increasingly commonplace, public water supply systems thus underwent expansion, both spatially and in terms of numbers of connections. By 1914, over three-quarters of the population of England and Wales had a piped water supply, two-thirds of which was supplied by municipal water authorities (Dracup 1973*d*). Over 96% of dwellings in London were connected to a water supply, compared with a figure of 17.5% for Paris (Goubert 1986: 198). The majority of cities drew water supplies from rural 'hinterlands', extending their ecological footprint into new river basins and across municipal boundaries (Dracup 1973*e*); Liverpool obtained water 68 miles away, Birmingham 74 miles away, and Manchester 106 miles away.[10] London's reach extended deep into Hertfordshire and Kent. For centuries an artisanal product, water had become an industrial product, and one of increasing national importance. Water was not only perceived as necessary for public health and hygiene, but also as a

[10] Cmd. 6513. *A National Water Policy* (1944).

source of hydroelectricity or cooling water for energy plants, and as a necessary requirement for large-scale industrial and urban growth.

The gradual consolidation of the legislative framework concerning water supply reflected this perception of water as a 'strategic' resource, integrally linked to public health, as implied by the delegation of responsibility for water resources to the Ministry of Health in the late nineteenth and early twentieth centuries. Subsequent Public Health Acts (1848, 1875, and 1878) detailed the general powers of local authorities with respect to water supply, empowering local authorities to 'undertake the necessary works to provide their district 'with a supply of water sufficient for public and private purposes' (Prusmann 1968: 64; Sheail 1983). This power was only to be exercised, however, in the absence of any commercial enterprise undertaking the water supply function to an adequate standard (Sleeman 1953). Many municipal water supplies were thus initiated under the auspices of the Health Acts, as opposed to the private Acts of Parliament authorizing private companies' operations, adding another layer of complexity to an already fragmented institutional framework. Amalgamation occurred gradually over the next three decades (Table 3.3). In contrast to the rapid expansion of water suppliers in the nineteenth century, the proportion of public to private suppliers remained almost unchanged between 1900 and 1945 (Hassan 1993), but the organization of public-utility services underwent significant changes during that period (Sleeman 1953). The extension of supplies to rural parishes, over 60% of which were without piped supplies in 1914, accelerated (Hassan 1998). By the end of WW II, there were over 1,000 water undertakers in England and Wales, of which 20% were statutory private companies, the remainder being individual and joint boards of local authorities (Prusmann 1968). The suppliers, however, varied greatly in size and coverage; 50% of the population was supplied by only 26 undertakings.[11] The complex mix of small, localized private and public service providers, in both water supply and sewerage services (the latter being almost entirely provided by local authorities) remained characteristic of the industry until its nationalization in 1974 (Prusmann 1968; Synnott 1985).

Reorganization and rationalization (1945–1974)

The complexity of the legislative framework, and the fragmented mix of public and private suppliers, sparked proposals for more centralized control of water resource planning (Sheail 1983). Water pollution was another motivation for calls for a central water authority. Although various pieces of

---

[11] For a more detailed discussion, see Appendix A of the White Paper on a National Water Policy (1943–4, Cmd. 6515).

legislation[12] addressed water pollution, enforcement was left in the hands of local authorities; responsible for sewage works, and thus one of the major polluters of inland waterways, local authorities had little incentive to address pollution (Smith 1972: 25).

The severe drought of the mid-1930s brought water resources issues into the political spotlight (Sheail 1983). Until then, Parliament had been reluctant to delegate further control over water resources, as 'the allocation of water resources affected property rights so closely that Parliament was particularly reluctant to delegate its regulatory powers over the industry to ministers and their officials' (Sheail 1983: 386). In light of powerful opposition to the 'bureaucratic encroachment on Parliamentary powers' (Sheail 1983: 394), encouragement of centralization and rationalization was sporadic and tentative (Prusmann 1968). Parliament chose to continue with the system of promoting private water supply bills, with the result that, with the exception of large cities, most supply systems were highly localized, and had limited storage capacity. The resulting insufficiency of water supply security was revealed during the 1934 drought, in response to which emergency legislation[13] was initiated, enabling much greater devolution of powers than was conventionally delegated to Ministers. With this precedent, inter-war initiatives to overhaul water resources and supply legislation gained further support (Sheail 1983); in 1945 the Water Act superseded the Waterworks Clauses Act of 1847, and initiated a process of rationalization of industry organization and structure.

This rationalization took place in a social and economic context that differed vastly from the pre-war era.[14] The reworking of the water supply system after 1945 must be viewed in the context of 'the extension of citizenship obligations during the war [which were] rewarded by the extension of social citizenship rights in the post-war period' (Hay 1995: 39). Although supplies in cities were nearly universal, 30% of the rural population did not have access to a piped water supply.[15] In the post-war period, the government proposed a significant Exchequer provision for water supply and sewerage,[16] in part to support the costs of providing water and sewerage infrastructure to New Towns (Sheail 1995). To enable the extension of networks in rural areas, local authorities were mandated to meet any shortfalls in the water supply service out of general rates, with the costs to be spread over the ratepayers of the entire district (not just the parish). The backlog of wartime underinvestment in housing also received attention; the demands entailed by government housing

---

[12] The Fisheries Acts (1861, 1865), the Rivers Pollution Prevention Act 1872 (based on the work of the Rivers Pollution Prevention Commission, from 1865 to 1874) (Smith 1972), and the Public Health Act of 1872, which abolished prescriptive rights to pollute and banned solid wastes from rivers.

[13] Water Supplies (Exceptional Shortage Orders) Act and Rural Water Supplies Act (both in 1934).

[14] In a report on water supplies published in the final year of the war, an *ad hoc* committee commented on the readiness in Parliament and public in general to consider constitutional changes that would strongly have been opposed beforehand (Sheail 1983).

[15] Ministry of Health (1944). A National Water Policy. Cmd 6513. London: Ministry of Health.

[16] £15m. (1949 prices).

programmes for water and sewerage facilities were provided for, at an estimated cost of £7.6 million (1949 prices) (Chick 1998).

During the post-war period, the water industry was regulated in a manner broadly consistent with the aims of other nationalized industries. At a macroeconomic level, the government applied 'Keynesian policies to support spatial cohesion of nation states and the full productive employment of spare regional capacity' (Swyngedouw 1989; Graham and Marvin 1995: 176). In common with practice in other utilities, pricing policy in water companies was based on price discrimination and promotion of sales in markets where demand was elastic, in order to spread fixed costs over output. The goal was to promote the fullest use of existing capacity, whilst attempting to avoid incurring the additional costs imposed by capacity constructed for peak load, particularly important in the water sector where a large proportion of capital expenditure goes to meet peak load capacity. Planning for growth to support domestic consumption and industrial production was thus a keystone of water policy.

In line with past practice, whereas meters for industrial users were widespread, domestic users were not metered. Water rates were based instead on the value of property ('rateable value'), thus allocating the costs of water according to (a rough indicator of) ability to pay. In common with other utilities in the post-war period, water supply services were used as instruments of social policy by granting price concessions to different classes of consumers (Sleeman 1953). This practice was underlain by an assumption that public utilities met basic needs, and supplied a service (provision of reliable, regular water supply), rather than a commodity (a specific amount of water at a specific time in a specific place) (Sleeman 1953). So great was this emphasis on water supply as a service that 'it came to be regarded as illegitimate to earn profits' (Sleeman 1953: 27) from the water business, and water companies, both public and private, were the only utility suppliers for whom a limit on earnings was imposed. In addition, a ceiling on water charges (as for other utility services) was specified by Parliament, for both statutory and local authority undertakings.

Additional controls were placed on private water companies, whose financial performance was much more tightly regulated than that of other private firms (Prusmann 1968). Share issues had to be authorized by Parliament (for both public and private suppliers); dividends on share capital were capped, first at 10% and later usually 7% or as low as 4% (Robson 1935; Sleeman 1953). The private statutory companies were thus, to a great extent, 'in the same financial position as publicly-owned undertakers' (Prusmann 1968: 79), subject to constraints similar to those placed upon their public sector counterparts, insofar as supervision of their activities was placed under direct ministerial control.

In many cases, rates were insufficient to cover capital maintenance as well as operating costs; often, neither private nor municipally owned companies generated sufficient funds for infrastructure renewal and water pollution prevention. For private companies, the allocation of profits was also regulated; companies could only retain a specified percentage of earnings, above which profits earned were applied to a reduction in the following year's water charges (Prusmann 1968). As the percentage retained was fairly small,[17] companies were forced to rely heavily on external sources of funding for both capital expenditure and refurbishment of fixed assets, pushing up the long-term costs of investment (Prusmann 1968). For public companies, parliamentary regulation ensured that water rates were 'at a level sufficient to cover capital costs and no more' (Prusmann 1968, 79), based on the 'political judgment [sic] that an adequate supply of water is a service to be provided at the lowest practicable cost on a non-commercial basis' (Prusmann 1968: 79). The resulting underinvestment had a significant, negative impact on river-water quality: 'There was a clear contradiction between the government's aims, as expressed in the pollution prevention and other water legislation it passed in this period, and its public expenditure decisions' (Hassan 1998: 112).

The recognition of increasing pollution problems, and the water industry's contribution to or inability to respond to those problems, led to a strengthening of the regulatory framework in the post-war period. The powers granted to the Ministry of Health by the 1945 Water Act were far broader than any legislation in the past, and included not only public supplies, but also conservation and water resources (Smith 1972). These powers were extended by the 1963 Water Resources Act, which further amalgamated governmental regulatory functions and administrative bodies,[18] creating River Authorities mandated to implement an abstraction licensing system over all of England and Wales (Smith 1972), thereby supplanting riparian rights, in practice, to a great degree.[19] Although nominally privately or municipally owned and locally run, the water industry was regulated along lines broadly consistent with the aims of other nationalized industries.

Municipalization thus entrenched patterns of water production established by private companies. Water supply was increasingly taken under corporate control (whether in the public or private sector), and an increasing proportion of water for human use was a networked, industrial product. Although some suggestions had been made for a move towards regional or catchment-based management, most water services remained highly localized and water management was largely characterized by a functional separation of responsibilities

---

[17] Companies were allowed to retain a reserve fund whose aggregate value was limited to 12.5% of total capital expenditure to date; annually, they could add to this reserve fund by no more than 1.23% of total capital expenditure to date. The amount carried forward on companies' profit and loss accounts was limited to the amount of the next year's dividend, plus interest payable, plus any arrears of dividend and outstanding interest (Prusmann 1968).

[18] Administrative bodies relating to matters as diverse as fisheries, navigation, land drainage, water resources planning, and water pollution were integrated.

[19] Licences must be obtained by all users except for individuals abstracting a limited amount (less than twenty cubic metres per 24-hour period) for domestic use or for agricultural purposes other than irrigation (Water Resources Act, 1991).

(between, for example, flood management, water supply, sewerage). Yet, as explored in the following section, these norms proved increasingly unworkable.

Corporatization and nationalization (1974–1989)

Nationalization entailed substantial reorganization of the industry, rejecting two cornerstones of the previous system: local government control, and functional separation of responsibilities (Dracup 1973e; Jordan, Richardson et al. 1977). Regional units of management were adopted. Within each of the 10 major catchments identified (Fig. 3.1), responsibility for source-to-mouth management was assigned to one 'Regional Water Authority' (RWA), consolidating statutory water undertakings together with local sewerage authorities in a basin-wide organization (Okun 1977). Nearly 200 statutory water undertakings, 29 river authorities, and 1,400 separate local authority sewerage services (Penning-Rowsell and Parker 1983; Synnott 1985) were consolidated into the 10 RWAs.

The 1974 nationalization of the water industry is anomalous within the utility sector. Unlike other utilities, for which municipalities and local private companies were 'no longer the operators of a large variety of small utility networks within their areas' (Graham and Marvin 1995: 175) after WW II, water remained fragmented, controlled largely by local authorities. Coal, steel, gas electricity, rail and canal transport were nationalized in the decade immediately following the war (Cairncross 1992). The nationalization programme was confined to fuel, power, transport, and communications, the most capital-intensive sectors of the economy. Nationalization of water thus occurred decades after other major utilities.

The Labour government's decision not to nationalize water in the immediate post-war period rested on several factors. Water shortages were not critical immediately following the war; for this reason water demanded less attention than, say, coal. For political reasons, the government was reluctant to tamper with water users' and property owners' rights. Municipal authorities also lobbied against nationalization, which would entail (as it eventually did in 1974) removal of a substantial degree of local control over water resources and supply (Hassan 1993).

Moreover, municipal control and greater ministerial authority meant, in effect, that the aims of nationalization could be achieved without direct central government control. With the majority of water services undertakings already under public ownership and increasingly reliant on central government funding, and through strict government regulation of those private companies still in operation, the central government was able to direct the strategic development of the industry. By the 1970s, the degree of central government control over water resources and supply had increased substantially, as had the pace and degree of industry rationalization, strongly encouraged by central

Fig. 3.1. Regional water authorities, England and Wales (1989).

government. From over 1,000 water undertakings in 1945, less than 200 undertakings were operating by the early 1970s, whilst covering a much larger territory and proportion of the population.

Despite rationalization, however, conflicts between water users were still common and water pollution incidents on the rise. Prior to 1974, water supply and sewage disposal had been the responsibility of public authorities; water reclamation was the exception. Sewerage authorities had an interest in disposing of effluent as inexpensively as possible; and, given, the fragmented, localized nature of water administration, not with water reuse in mind (Jordan, Richardson et al. 1977). By the late 1960s, an estimated 60% of all sewerage works failed to meet Royal Commission standards established at the turn of the century (Smith 1972: 178). Water-quality problems were compounded by the fact that throughout much of England lowland rivers played (and continue to play) both an important effluent disposal function, and a water supply function. In light of forecast water shortages,[20] and with a declining number of clean upland water sources, the prospect of widespread reuse (i.e. effluent recycling) seemed likely; the need to integrate decision-making, on both a spatial and functional basis, took on greater importance (Penning-Rowsell and Parker 1983).

Following nationalization, the management of water and sewerage was regionalized. The Regional Water Authorities operated as vertically integrated monopolies within watersheds, based upon the principle of Integrated River Basin Management (IRBM). A central principle of IBRM is that water and sewerage management should be unified and organized geographically, on the basis of river basins. IBRM is based upon the understanding that water, as a multipurpose flow resource within hydrologically interconnected systems, is required simultaneously, or sequentially to fulfil several distinct functions (Rees 1989: 29); this is particularly important in densely populated river basins largely dependent upon surface sources for water supply. Integrated management can thus minimize conflicts between uses, as well as capture the economies of scale and agglomeration available when a total regional water cycle is managed as a unit. In its application in England and Wales, IRBM required that water authorities both operate (as potential polluter or 'poacher') and also regulate (as 'gamekeeper') water supply and sanitation systems. RWAs thus monitored, policed, and prosecuted themselves (and the remaining private water companies) on water quality (Kinnersley 1994). The management of 'clean' and 'dirty' water functions was thus unified under the control of the RWAs.

Administrative control was also regionalized (rather than centralized) under amalgamated bodies, on whose boards local government had only minority representation (removed altogether under the provisions of the 1983 Water Act), reflecting a managerialist ethos and prioritization of technical efficiency, rather than public accountability (Jordan, Richardson et al. 1977). Prioritization of administrative and economic efficiency over local democratic control reflected more widespread administrative reform, removing those functions from local authorities that could be more rationally managed at larger scales (Penning-Rowsell and Parker 1983). Industry structure was not, however, completely rationalized; 28 statutory private water companies remained, operating as licensed agents of the RWAs, supplying approximately 20% of total consumption (Fig. 3.2).

[20] These inflated forecasts were based on a simple extrapolation from the rapidly increasing industrial demand of the 1950s and 1960s.

Building the Networks

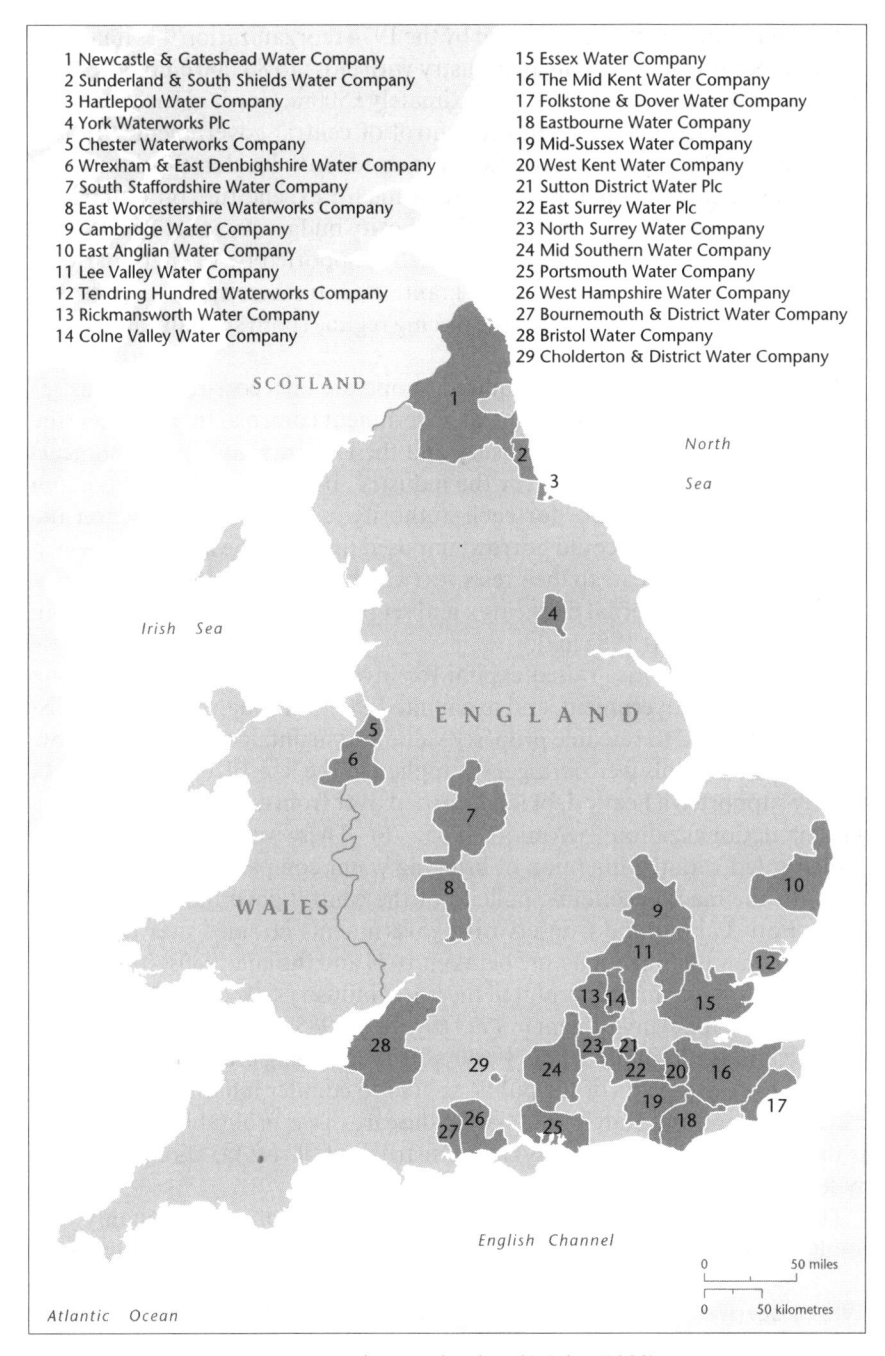

1 Newcastle & Gateshead Water Company
2 Sunderland & South Shields Water Company
3 Hartlepool Water Company
4 York Waterworks Plc
5 Chester Waterworks Company
6 Wrexham & East Denbighshire Water Company
7 South Staffordshire Water Company
8 East Worcestershire Waterworks Company
9 Cambridge Water Company
10 East Anglian Water Company
11 Lee Valley Water Company
12 Tendring Hundred Waterworks Company
13 Rickmansworth Water Company
14 Colne Valley Water Company
15 Essex Water Company
16 The Mid Kent Water Company
17 Folkstone & Dover Water Company
18 Eastbourne Water Company
19 Mid-Sussex Water Company
20 West Kent Water Company
21 Sutton District Water Plc
22 East Surrey Water Plc
23 North Surrey Water Company
24 Mid Southern Water Company
25 Portsmouth Water Company
26 West Hampshire Water Company
27 Bournemouth & District Water Company
28 Bristol Water Company
29 Cholderton & District Water Company

Fig. 3.2. Statutory water companies, England and Wales (1989).

A third important change initiated by the 1974 reorganization was financial, as water sector investment, in an industry with close to 80,000 employees and an annual investment budget of approximately £500m. (Jordan, Richardson *et al.* 1977), now fell under the direct control of central government. Prior to 1974, income from water rates was not 'ring-fenced'; local authorities could determine whether income should be set aside for expenditure on water services, or absorbed into the general local authority budget (Synnott 1985). Until 1974, the cost of water services was partially supported by the national government through the rate-support grant, and further by cross-subsidies between consumers embedded in the pricing regime (both spatially and across socio-economic groups).

After 1974, the RWAs were required to operate on a cost-recovery basis,[21] and were subject to additional central government controls. In particular, the government (in practice, the Treasury and the Department of Environment) set specific fiscal requirements for the industry. It determined the maximum annual capital expenditure for each authority; set an annual limit on the amount each authority could borrow; imposed limits on the amount to which the RWAs could allocate to their reserves (which in turn restricted the proportion of internal to external financing); and set performance aims for operating expenditure (Synnott 1985: 67).

The water authorities raised capital for investment mainly by borrowing from the central government, and to a limited extent through customers' bills, which were linked to rateable property value and maintained at relatively low rates. Fiscal controls were stringently applied in the late 1970s, as the RWAs already supported a heavy debt load, carried over from municipal debts at the time of nationalization[22] (Synnott 1985: 76). These various controls were implemented with the intention of bringing water company expenditure into line with the macro-economic policies of the central government (Meredith 1992). Both Labour and Conservative governments presided over decreased spending on water infrastructure between 1974 and the late 1980s. In 1982, for example, the government permitted the water industry to spend only half of its total annual capital investment in 1974 (Synnott 1985; Ward 1997: 94). RWAs thus became agents of fiscal policy (Synnott 1985), being used, for example (as were other components of the public sector) to counter inflation by reducing costs (Redwood and Hatch 1982), or (in the case of the profitable RWAs, such as the Thames Water Authority) as an instrument of soft taxation, as argued by some RWA chairmen (Harper 1988).

This underinvestment, and declining infrastructure and water quality that resulted, stood in sharp contrast to government rhetoric over nationalized

---

[21] Section 29 (1) of the 1973 Water Act states that: 'It shall be the duty of every Water Authority to discharge their functions so as to secure that taking one year with another their revenue is not less than sufficient to meet their total outgoings properly chargeable to revenue accounts.'

[22] By the early 1980s, nearly 50% of the annual RWA revenue budget was taken up by interest and depreciation, a significant proportion of which arose from the inherited debt (Synnott 1985).

water provision. To a greater degree than any other public good the provision of water was thought to be characterized by the market failures inherent in natural monopoly, as evidenced by the performance of private water companies in England in the nineteenth century (Millward 1989). The underlying definition of public goods was of non-substitutable, essential goods, collectively consumed, necessary for both production and reproduction. Universal access to public goods, including water, was regarded as a social right and a necessary element of the Fordist social contract between state, citizen, and capital (Graham and Marvin 1995). Public ownership of utilities was underpinned by a Keynesian model of social welfare, proponents of which argued that state provision of merit goods was in the general economic and social interest (Ernst 1994; Graham 1997). Cross subsidies—between regions, socio-economic classes, and industries—were accepted and even encouraged.

This financial strategy was simultaneously a political strategy. In the UK, investment in the utility sector was used to bolster state legitimacy amongst citizens, thus ensuring political stability with a 'peace dividend' after the near-full mobilization of the electorate during WW II (Hay 1994*a*; Hay 1996). Utility investment thus answered to the twin imperatives of the politico-economic settlement and politico-normative settlements. This had important financial consequences. In the water sector, by the 1970s, levels of debt and interest payments grew beyond what the government was willing to support through public sector borrowing. This shortfall increased throughout the 1970s and on into the 1980s, when the Thatcher government's monetarist policies began to be implemented.

The effect on the water industry, as with other areas of the public sector, was a sharp reduction in capital expenditure allowances. With little political acceptance of dramatic increases in water bills, the water authorities faced a cash crisis. Responsible for water quality yet unwilling to self-prosecute and with limited scope for investment, both water and infrastructure quality deteriorated significantly across England and Wales (Pearce 1982; Kinnersley 1988; 1992). Government encouragement to RWAs 'to view themselves as suppliers of related, but separate commodities, rather than as strategists for the delivery of an integrated service' (Synnott 1985: 169), leading to a functional separation of responsibilities within the authorities, further discouraged consideration of the interrelationship between 'clean' and 'dirty' water. Investment decreased dramatically, bills were not permitted to rise quickly enough to cover costs, and the cash flow of some authorities diminished severely. Water authorities cut back on investment in new waste water treatment works, and maintenance and repair of existing 'dirty' water systems. Throughout the 1980s, under-investment in 'dirty water' systems was the norm (Rees 1989). In effect, as the government ran out of options to handle the financial costs of the politico-normative settlement of universally provided water, it enacted policies which forced the water authorities to externalize those costs on to, and beyond, the water network.

As a result, pollution incidents rose sharply. The relaxation of consent conditions for industry (and for water authorities own sewerage works) temporarily masked declining water quality (Pearce 1982). Yet by the early 1980s awareness of the water industry's 'crisis' situation began to grow, as river quality began declining after a measure of improvement in earlier decades. Britain earned the media label the 'dirty man of Europe', dumping more sewage sludge in the sea than any other country in the world, and failing to improve the condition of over 3,000 kilometres of river too polluted—with industrial effluent as well as sewage—for any fish to survive (Pearce 1982; Rose 1991). At the time of nationalization, the government had passed the 1974 Control of Pollution Act, aimed at controlling pollution in land, air, and water; the provisions on water were still unimplemented in 1982 (Pearce 1982).

The perceived crisis in the water industry in the 1980s was thus simultaneously fiscal and environmental. These were not, however, new or sudden developments. Infrastructure and water quality had been declining since the mid-1970s, after a period of higher investment prior to nationalization in the early 1970s (Pearce 1982). The twinned quality–financial crises provided a powerful vehicle for legitimating change in the water industry. From the neoliberal perspective, the crises were evidence of the inefficiency of public water companies, and a justification for privatization (Vickers and Yarrow 1988). Failure in the water management network provided the preconditions for the political mobilization of perceptions of crisis; the 'crises' in water management were in part the outcome of a political mediation of symptoms of water management failure.

This discursive depiction of 'crisis' had a pragmatic edge. After decades of underinvestment in the water industry, the government was faced with a massive capital investment requirement following the introduction of stringent European Union water quality regulations concerning the quality of drinking water and bathing waters (Walker 1983). After decades of deliberate non-compliance with EU directives on the part of the government, the decision of the European Union to prosecute Britain for non-compliance in the mid-1980s was politically decisive (Hassan 1998); a major capital investment programme on water quality was unavoidable. The projected 10-year capital investment requirement was estimated at between £24 and £30bn. in 1989 prices (Ernst 1994: 77; Hassan 1998: 167). Committed to monetarist economic policies (Foster 1992), the government was unwilling to fund the required investment either through increasing taxes or raising the Public Sector Borrowing Requirement (PSBR). In effect, the government was having increasing difficulty 'balancing ideological objectives (RWA commercial autonomy) with fiscal imperatives (increasing RWA cash flow to lessen PSBR impact)' (Synnott 1985: 125). Privatizing the industry, it was thought, would allow the government to raise revenue, and more importantly to avoid future capital expenditure burdens.

Moreover, water was no longer such a strategic resource. As industrial consumption declined relative to domestic consumption, levels of connection to the water networks, and water usage, were less useful surrogate indicators for national economic performance than they had been in the past (Sleeman 1953). Total water consumption declined in the mid-1970s and subsequent decades (Fig. 3.3) as industrial consumption fell,[23] contrary to the predictions of the Water Resources Board.[24] Not only did overall industrial consumption decrease, so too did the proportion of water-intensive industries. Greater use of water-efficiency measures also contributed to the fall in demand (WSA 1987; 1998). The change in demand was also qualitative, with the simultaneous emergence, and predicted growth of industries requiring smaller amounts of very clean water (for example the high-tech sector) rather than large amounts of poorer quality water (Scottish and Westminster Communications 1992).

This factor was an additional reason for increased concern over declining water quality in the 1970s and 1980s. As its monetarist policies were non-negotiable, the Conservative government held to its decision to prioritize the PSBR over water quality improvements and pollution prevention. The

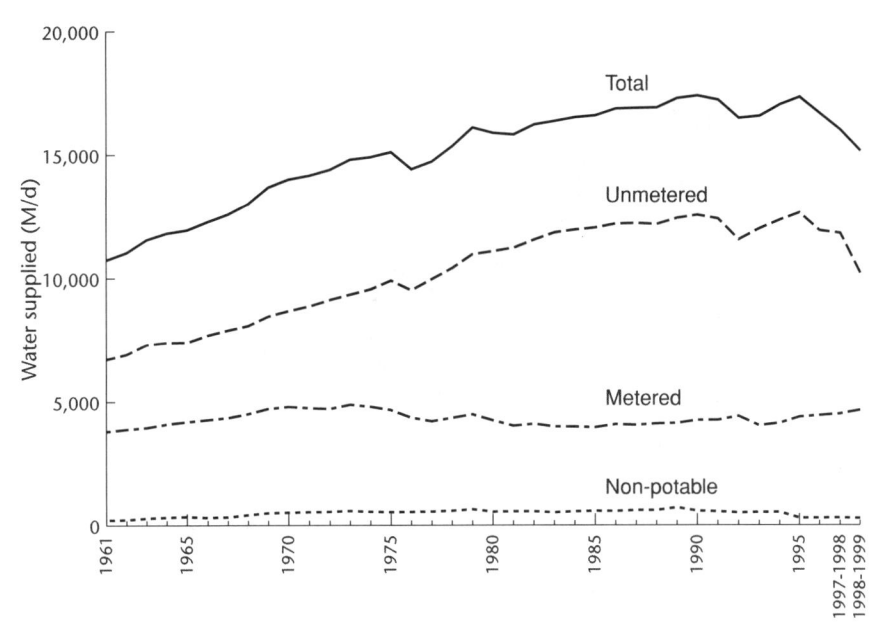

Fig. 3.3. Public water supplied, England and Wales (1961–1999).
(*Data source*: WSA (various years))

---

[23] As the majority of industrial users are metered, and the majority of domestic users un-metered, metered consumption is taken as a proxy for industrial consumption.

[24] The Water Resources Board sponsored the construction of many resource developments that lay idle in subsequent decades, in part because although it correctly forecast rising domestic consumption, it failed to predict falling industrial consumption.

government was left with two options: either devolve the costs of water quality improvement onto consumers through massive increases in water bills; or continue to allow water quality to deteriorate. Whilst allowing limited increases in water bills, it attempted to commercialize water management, thereby increasing efficiency, as the means of sourcing the extra capital required. The Water Act of 1983 was thus predicated on the assumption that consumers were best served 'by an efficiently run operation, providing the requisite service at least cost' (MMC 1981: 264). This Act initiated many of the fiscal changes now attributed to privatization, and marked the transformation of the water industry in Britain 'from a public service to a business organization' (Penning-Rowsell and Parker 1983: 170). Cost-benefit analysis was introduced to the industry, and attempts were made to introduce long run marginal cost pricing (Synnott 1985: 70). Efficiency measures were introduced; for example, job shedding in the industry dates from the early 1980s with employment levels peaking in the early 1980s and reducing substantially before privatization in 1989 (Fig. 3.4) (O'Connell-

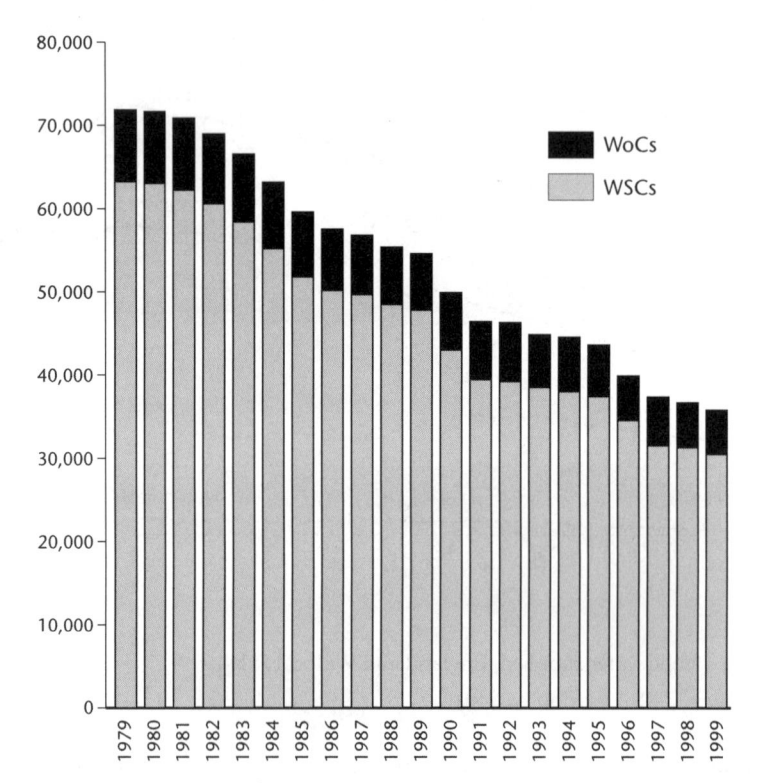

Fig. 3.4. Employment levels (full-time equivalent), English and Welsh water industry (1977–1999).

(*Data source*: WSA (various years))

Davidson 1993). By the late 1980s, the nationalized industries were best characterized as 'publicly regulated private monopolies operating on modified market principles' (Hay 1996: 53).

This set of strategies did not, however, result in significant change in water quality, which continued to decline throughout the 1980s. New water quality requirements imposed by the European Union necessarily entailed large capital expenditure to improve water quality (Kinnersley 1994; Ofwat 1992*a*). The EU's willingness to enforce these regulations, as demonstrated by its initiation of legal proceedings against the British government in the late 1980s (and again in the early 1990s) for failing to implement two of its water quality directives,[25] foreclosed the possibility of continuing the practice of underinvesting in water quality treatment infrastructure.[26] With its broader programme of privatization of public sector enterprises well underway, the government issued a discussion paper on water privatization in 1985. After an initial announcement, with the realization that water would require a more complex regulatory framework than other network industries, water industry privatization was put on hold. After a fitful and lengthy policy process the government decided to move ahead with the privatization of the water authorities in 1989 (Kinnersley 1994).

Privatization of the water industry

In December 1989, despite polls indicating that three-quarters of voters disapproved of water privatization (Saunders and Harris 1994), the 10 regional water authorities were floated on the stock exchange. The boundaries of water supply areas were unchanged by the transfer to private ownership. In their existing regions of operation, the WaSCs were appointed as vertically integrated regional monopolies, providing the entire cycle of services to their customers, from extraction of raw water, delivery of processed water, and collection, treatment, and discharge of waste water. The smaller water supply companies were appointed to provide water services to their previous customer base.

The divestiture of the water authorities was part of a broader government strategy of revenue acquisition from the one-time sale of assets. Throughout the 1980s, privatization proceeds in effect disguised the fact that public expenditure had risen (Foster 1992) due to the national accounting convention of allocating

---

[25] The two directives in question specified standards for the quality of water for human consumption (80/778/EEC), and for bathing waters and beaches (76/160/EEC) (Brown 1992*b*).

[26] The Treasury's refusal to consider bond issues (commonly used by publicly owned water companies in the United States) meant that the water companies could not access private capital markets while under public ownership. In the British system of National Accounts, the PSBR calculation includes all borrowing by public corporations, as well as by central and local government (see s. 21.14 of ONS 1998). The high capital expenditure requirements of water companies conflicted with the government's desire to hold down the PSBR, and thus borrowing by individual water authorities was tightly constrained by the central government from the early 1980s onwards.

privatization proceeds as 'negative public expenditure' (Foster 1992: 116). By the late 1980s, revenue from asset sales was substantial enough to allow public debt repayment (Curwen 1994). This revenue is a one-time gain and should have been offset against the future stream of profits foregone by the Government through the sale of the nationalized industries, to determine whether the net effect on the PSBR was positive or negative (Curwen 1994). However, the system of National Accounts classifies revenue from the one-time sale of assets as a simple payment, such that privatization proceeds reduced and indeed helped create a negative PSBR for several years during the mid-1980s. Privatization was thus simultaneously a fiscal and a political strategy—although not a successful strategy in fiscal terms in the case of the water industry.

As stated by the Department of the Environment in 1985, the objectives of the privatization programme were to: promote competition and enterprise; reduce the size of the public sector; involve staff in companies; spread share ownership; and free enterprise from state controls (DOE 1986). In the case of the water sector, an additional objective was paramount: to avoid responsibility for the large capital expenditure burden imposed by new European legislation requirements on bathing waters, beaches, and drinking water quality. A central strategy of the government was thus to disengage financially from water production.

Although divestiture was a key strategy for revenue gain on the part of the government's privatization programme, in the case of the water industry, the government's loss totalled £1.2bn. The £6.4bn. 'green dowry' consisting of £4.9bn. in debt write-off and a £1.5bn. cash injection (1989 prices) was not completely recovered (Ofwat 1997a: 10, n. 3), as the 10 WaSCs sold for only £5.2bn. Moreover, the costs of privatization, including underwriting, advertising, and free and discounted shares to employees totalled £146m. (Curwen 1994), of which £36m. alone was spent on advertising, the highest figure by far for any privatization. In addition, the water industry was granted £7.7bn. in tax concessions, and given the majority share of the industry's pension fund, creating a shortfall of up to £660m. in the fund of those employees of the water companies that remained with the public sector in the regulatory body then known as the National Rivers Authority (now the Environment Agency) (Hassan 1998; National Audit Office 1998; Shaoul 2000a). Although the lack of finance for public-sector investment in water provision is a key justification for privatization, in this case, privatization itself proved to be loss-making for the government.

The 'British model' of regulation

The post-privatization regulatory framework is characterized by a functional separation between economic, product quality and environmental regulation. The environmental regulator—first known as the National Rivers Authority, which was incorporated into the Environment Agency in 1996—was hived off from the water industry in 1989, remaining in the public sector. In addition to water resources and abstraction licensing, the EA is responsible for pollution control, fisheries and flood defence, as well as with the management of water resources, including resource planning and the licensing and regulation of water abstractions. Drinking water quality monitoring (through the setting and enforcement, by prosecution if necessary, of minimum water quality standards) is the responsibility of a second product quality regulator, the Drinking Water Inspectorate (DWI). Control of water resources abstractions and responsibility for water quality standards were thus retained by state regulatory agencies.

With the persistence of regional monopoly and in the absence of competition, regulatory control on abuse of monopoly power was necessary; hence the creation of a new economic regulatory framework. Economic regulation is the responsibility of a third regulator, the Office of Water Services (Ofwat), which sets limits on price increases charged to consumers, and is charged with the duty of ensuring that the private water companies have adequate finances to carry out their core functions. Characterized by price-cap regulation and simulated or 'yardstick' competition (Helm and Yarrow 1988; Littlechild 1988; Armstrong, Cowan, and Vickers 1994), the new style of economic regulation was designed to encourage increased efficiency and distance the management of water companies from hitherto frequent political interference. Charged with the duty to ensure that the appointed companies are able to properly finance the carrying out of their functions, Ofwat sets the 'price cap' limit on increases in the consumer tariff basket every five years (Armstrong, Cowan, and Vickers 1994). Facilitating competition and protecting the interests of consumers are secondary duties (OXERA 1994*a*). Both in terms of the information which it is empowered to gather from companies, and its power to approve company investment plans, Ofwat is the more powerful of the three core regulators (Helm and Rajah 1994), and is the key actor in the commercialization of the water supply industry, particularly through the duty to increase competition in the industry, where feasible.

In addition to the three regulatory bodies, the Department of the Environment, Food and Rural Affairs (DEFRA) plays a central role, setting standards and formulating water policy. DEFRA also plays a role in regulation; its role is not as clearly defined by legislation as that of the three primary regulators, in keeping with the British tradition of informal, non-legalistic, and discretionary regulation. Although originally expected to have an arm's-length role, DEFRA assumed greater importance when it chaired the 'quadripartite' process created in 1994 when the Periodic Review process, originally to be held once every decade, was initiated only five years after privatization. In addition to DEFRA, the remit of a series of government departments and regulatory bodies overlaps with the activities of the water sector, and the industry is subject to intervention by a complex network of governmental actors at a variety of scales (Fig. 3.5, Table 3.4). Most importantly, the European Union plays a key role in legislation and standard-setting, formerly through issue-based directives, and now through the comprehensive Water Framework Directive. Global localization, or the simultaneous scaling-up and scaling-down of regulation, is evident in the partial 'hollowing out of the state' with the reallocation of responsibility for setting some water-quality standards to the European Union, whilst implementation and enforcement have been devolved to the regional level (Buller 1996; Hassan 1995; Richardson 1994). The government has divested control over the financial management of the water industry, but maintained control over regulatory functions concerning health, environment, and resource management, functionally separating the regulation of 'market' and 'consumer' interests.

Where the inevitable contradictions between the politico-economic and politico-normative settlements were once addressed within a few government departments, the creation of quango regulators, and the multiplication of regulatory bodies has diffused responsibility and delegated accountability.

Non-Governmental Bodies Attempt to Shape Policy

- NGO's representing landowners and rural Business (National Farmers Union, Country Land and Business Association)
- Environmental NGOs (Greenpeace, Council for the Preservation of Rural England, Friends of the Earth, Royal Society for the Protection of Birds)
- NGOs lobbying on behalf of specific social groups (Age Concern, Help the Aged, Save the Children, National Council for One Parent Families, Child Poverty Action Group)
- NGOs focused on UK Water Industry (Waterwatch, Public Utilities Access Forum, National Campaign for Water Justice)
- Independent Consumer Groups (Consumers Association, National Association of Citizens Advice Bureaux, National Federation of Consumer Groups)
- Unions representing workers in the Water Industry (Unison, Public Services International)
- Private Water Companies
- Industry Representatives and Trade Bodies (Water UK, Society of British Water Industries, Chartered Institute of Water and Environmental Managers, Confederation of British Industry, Chemical Industries Association)

Committees Appointed by Goverment Represent the Public in the Process of Making and Implementing Policy

- Sustainable Development Commission
- Environmental Agency Advisory Committees (RFERAC, REPAC, AAC's)
- Ofwat National Customer Council (ONCC)
- Water Panel of the Competition Commission

Appointed Regulators Produce Regulations & Police Policy

- Environmental Agency (EA)
- Drinking Water Inspectorate (DWI)
- Office of Water Services (Ofwat)
- Competition Commission (CC)

Government Creates Laws & Policy

- The Department of the Environment, Food and Rural Affairs (DEFRA)
- Ad Hoc initiatives by Ministers (e.g. Water Summit)

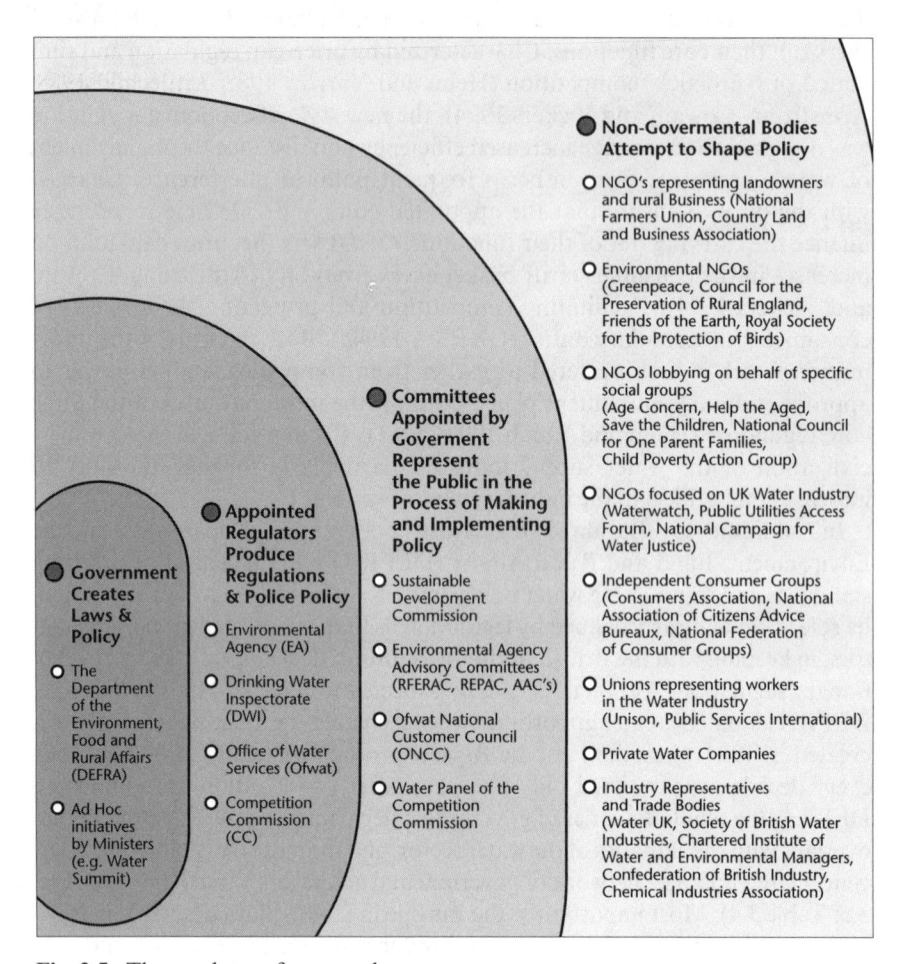

Fig. 3.5. The regulatory framework.

(Adapted with kind permission from Page and Swyngedouw (2002))

Table 3.4. Water regulation in the UK

Scale	Regulator	Role	Issues
European Union	EC	Legislation, standard setting	Comprehensive: water resources, wastewater, water quality, pricing, environmental
National	Environment Agency	Environmental regulator	Regulatory (water resources, river water quality) and operational (flood control, fisheries, navigation)
	Office of Water Services (OfWat)	Economic regulator	Grant licence to supply; supervise competition; set price caps; monitor service quality; handle complaints
	Department of Environment, Food and Rural Affairs	Standard setting, drafting of legislation, water policy, appoints regulators, special permits (e.g. drought orders)	Comprehensive
	Drinking Water Inspectorate	Water quality regulator	Drinking water quality monitoring and enforcement
	Competition Commission	Reporting; Appeals	Monopolies, mergers and the economic regulation of utility companies; company mergers; price cap limits
Regional	Environment Agency	Tribunal on regulatory decisions	Water resources (water abstraction licences); environmental water quality (discharge consents; monitoring)
	Ofwat Customer Service Committee	Handles water company customer complaints; represents consumers in Ofwat policy-making process	Standards of service; compensation; customer complaints

The implications for environmental management of this diffused accountability, and separation of regulatory responsibility for capital expenditure and environmental planning (in the hands of the environmental regulator, but effectively controlled by the economic regulator given his authority over 'environmental spend'), are detailed in Chapter 5. Chapter 6 considers the implications for consumers of the functional separation and specialization of regulatory roles, which entailed an attempt divest the post-privatization regulatory framework (and, in particular, the economic regulatory function) of 'social policy' considerations. In Chapter 7, water supply companies are the focus; we consider the long-term viability of the 'British' model of vertically integrated, equity-financed private monopoly suppliers of water services.

Conclusions

The extension of water services to the majority of the population under the auspices of municipal governments was characterized by an emphasis on the provision of water as a service; the goal of universal provision; equity as the paramount policy principle; and planning for growth. The regulatory ordering of water supply from the late nineteenth century onwards followed the logic similar to that of other utilities: municipalization; rationalization; and nationalization, accompanied by a gradual loss of accountability to local levels of government. The contemporary organization of water supply production at the scale of river basins is linked to this regulatory reordering of water management. The early archipelago of networks concentrated in urban areas, with great diversity in tariffs, levels of connections, spatial extent of networks, and quality of service, was gradually displaced by (partially) regionally integrated systems operated according to a supply-side logic. The consolidation of the 'state hydraulic' paradigm during the mid-twentieth century was characterized by rationalization, decreasing local government control over water supply functions and consolidation of water supply management into increasingly large territorial units, culminating in the large river basins controlled by the Regional Water Authorities.

Given the size and hydrological interconnectedness of these regional monopolies, individual water companies increasingly bore the costs of the negative externalities they themselves produced; in the context of more stringent water quality regulations, those costs increased sharply. This decrease in quality was perceived as increasingly unacceptable given heightened environmental awareness in the 1970s and 1980s, and increasingly stringent EU water quality legislation. Water managers (as in other sectors) were confronted with one of the central dilemmas of the Keynesian state: their continued legitimacy was dependent upon the satisfaction of expectations sanctioned by the state, but provision of which threatened to undermine economic competitiveness and environmental quality (Hay 1996). The exhaustion of options for dealing with this dilemma at a macroeconomic level gave rise to the crisis of Britain's 'monopolistic' or 'administered' mode of regulation in the late 1970s.

The gradual shift towards 'market environmentalism' began at this time, as attempts to stave off macroeconomic crisis by starving the water sector of capital, resulted in a twinned crisis of finance and water quality to which the pragmatic response was commercialization. Commercialization was characterized by tighter financial controls, a reduction in investment, sustained price increases above inflation, and an increasing emphasis on economic, as distinct from technical performance indicators. These changes were due in large part to the macroeconomic crisis, which was mediated through the water sector via eco-regulatory mechanisms (such as the lowering of water quality standards and declining investment).

Commercialization, in enabling an (unsustainable) 'ecological fix' to be implemented, served as a mechanism of crisis displacement by the state, but in turn served to create a twin ecological–financial crisis in the water supply sector. Privatization offered an apparent fiscal, ideological, and political resolution to this crisis. Privatization allowed the government to reduce the PSBR, and deliver another blow to the power of the unions (by encouraging employee share ownership, as well as moving the water industry into the private sector), whilst simultaneously distancing itself from public displeasure over the rate increases necessary to fund required capital expenditure (Rees 1989; Ogden 1991; Ferner and Colling 1993; Saunders and Harris 1994). The regulatory framework created post-privatization was predicated on an implicit assumption of 'state failure' (rather than market failure) in water supply, and was designed to deepen the commercialization of the water supply sector, notably through incentive-based price cap regulation designed to encourage productive efficiency gains and (later) the introduction of competition. The consolidation of the 'market environmentalist' paradigm via the ongoing commercialization of the water industry is analysed in detail in Chapter 4.

4

Commercializing water supply

The progressive commercialization of the water industry is the subject of this chapter. Recall the definition of commercialization stated in Chapter 1: the introduction of markets and market-simulating techniques as allocation and decision-making mechanisms, and the application of commercial practices (for example, cost–benefit analysis) and commercial goals (for example, profit maximization). In general, the successful commercialization of a service has four key prerequisites: the creation of an effective demand (not applicable in the case of water supply, given that it is partially non-substitutable resource essential for life); the conversion of a good or a service into a saleable commodity; the conversion of the workforce into one oriented towards profits; and the underwriting of risk by the state (Leys 2001). Full commercialization requires the correct valuation of water and all of the externalities associated with its production, consumption, and disposal, in order to enable the correct pricing of water and communication of that price to consumers.

But what is the correct price of a glass of tap water? The true value of a stream? The proper pricing of water, and value of environmental 'externalities' associated with its abstraction, production, and disposal, have received a great deal of attention in the water industry since privatization. This is in part because the methodologies employed to assign monetary values to nature are difficult and disputed. It is also because the valuation of nature is a necessary but controversial step in the progressive commercialization of the water industry. For although privatization entailed a decisive change in ownership, regulatory structure, and financing mechanisms, it did not entail the immediate introduction of market mechanisms into water allocation decision-making, nor the conversion of water into a commodity.

To argue that water is not a commodity, despite the fact that it has a price and is delivered by private companies, may seem at first disingenuous. But consider the definition of a commodity as a standardized good or service, with interchangeable units, sold at a price determined through market exchange. In England and Wales, water has a price, but is not standardized and its units are not interchangeable; water supply networks are not integrated even within company supply areas, and quality and other parameters vary significantly from one area to the next, posing complex (and some would argue insurmountable) technical difficulties for the creation of a national grid. Hence water is not traded in bulk—there is no national, and few truly regional water grids in the UK, and water transfers between companies are limited in volume. Nor is the price of water determined by market exchange, that is, relative to supply and demand as determined through exchange contracts between buyers and sellers. When water companies pay for the abstraction of water, a charge is set; but (until recently) this covered solely the cost of administering the abstraction licence, rather than the value of the water and the externalities which water abstraction generates. When consumers pay for the water they use, charges for the majority of domestic users, who do not have meters, are set relative to the property value of their home. For the majority of consumers, then, water is not priced per unit volume; its price thus bears no relationship to demand. The principles underlying charging mechanisms are that of water supply as a service. And externalities generated by water abstraction and disposal are not incorporated into its price.

Networked water supply in England and Wales is a 'quasi-commodity', or only partially commodified. The difficulty which follows from the economist's perspective, as explored in detail below, is that the production of water will be less than optimal—the market will not function as an efficiency-maximizing social institution for the allocation of resources. Post-privatization, then, much regulatory attention and ingenuity has been dedicated towards commercializing the water supply industry, a far from straightforward process. Post-privatization, the regulatory and legislative framework envisaged the introduction of widespread metering, the unravelling of cross subsidies, the introduction of cost-reflective pricing, and implementation of competition in the water supply industry. These measures have been put in place, but much more slowly and subject to many more constraints than anticipated in 1989. The commercialization of the water supply industry has been fraught with difficulty, and in some cases controversy.

It is important to note that commercialization is a process enacted not solely by the water companies and regulators, but also by government, creating required institutional frameworks for water resources and supply management, and establishing the necessary governance structures and relationships between citizens, companies, governments, and water environments. As such, the process of commercialization is simultaneously environmental, social, economic, and political. As the means by which market environmentalism is operationalized—promising improved economic efficiency and water conservation, greater economic equity and environmental protection—this process deserves careful scrutiny. The implications of the continuing commercialization of water supply for water management, consumers, and the water supply industry will be considered in detail in subsequent chapters.

This chapter examines three of the most important aspects of commercialization more closely: valuation; the creation of price signals (via metering, with the goal of facilitating allocation of water to its highest value uses); and liberalization (the introduction of competition). The chapter argues that each creates trade-offs, creating dilemmas for regulators attempting to introduce a greater degree of commercialization into the water supply industry. To some degree, the government has responded to these trade-offs by retreating from its earlier commitment to full commercialization of the water industry; the analysis of reasons for and implications of this 'retreat from commercialization' is begun here, and continued in subsequent chapters.

Valuing water

To fully commercialize water supply is more complicated than might first appear. Converting water into a commodity requires: cost-reflective pricing; a network or other mode of supplying water in standardized units; and volumetric metering of all consumers so that they may receive and respond to price 'signals'. Cost-reflective pricing, for example, requires pricing which incorporates externalities, the cost of which is difficult to measure. If prices were to accurately reflect costs, then each consumer must be charged a price related to the costs they impose on the system, implying temporal and spatial variation in pricing—water should cost more during a dry period in the summer; rural consumers would pay more than urban consumers. The spatial and temporal disaggregation of pricing and the valuation of externalities are technically difficult and politically contentious.

The term 'externalities' was first used to address pollution problems in Pigou's system of taxes and subsidies intended to correct for social costs excluded from private decision-making (Pigou 1920; Eckersley 1995). Coase's challenge to the Pigovian approach identified the absence of markets and associated property rights as the cause of externalities, but retained the notion of their exclusion from the realm of the market (Coase 1960; Helm and Pearce 1992). The remedy for externalities is better costing of environmental goods, and incorporation of these valuations into measures of economic performance that feed into policy decision-making, i.e. better 'price signalling' (Dasgupta 1990). Some economists argue that these valuations can only work in the presence of more robust property rights, and regulation designed to increase private incentives for obtaining proper scientific information about resources (Dasgupta 1990). Others argue that fewer regulations and greater competition are a more efficient institutional mix (Robinson 1997). But they all agree on the need for valuation of water—fully commodified—if incorporation as a commodity into market exchange is to prove workable.

Accordingly, there has been a concerted effort since privatization on the part of regulators and the government to commercialize water. This has been justified by the belief that treating water as a quasi-commodity leads to inefficient management, which not only leads to higher costs and thus higher prices, but also leads to inefficient use of water. Whereas these effects might have been tolerated or overlooked under the 'state hydraulic paradigm', under which efficiency was often treated as a secondary objective, efficiency is increasingly asserted to be the primary objective of water management under the market environmentalist paradigm. The notion of economic efficiency employed here implies that maximizing efficiency entails the allocation of water its highest value uses. Efficiency is understood to be equivalent to Pareto-optimality (a concept derived from welfare economics): 'a situation in which no feasible change can lower anyone's welfare without lowering that of someone else' (Black 1997). In the language of welfare economics, the Pareto-optimal path for the water industry is that in which the marginal benefit of the next increment of water supplied would equal the marginal cost of supplying that increment. The more efficient the water management 'solution'—here understood as the allocation of water to its highest value uses—the higher society's 'welfare'.

The ascendancy of efficiency

The prioritization of efficiency, and the underlying definition of welfare maximization as efficiency maximization, is relatively recent in water supply management in England and Wales. Neither operational nor economic efficiency were paramount performance criteria for the water industry before its nationalization in 1974. Rather, universal access and security of supply were prioritized; welfare was implicitly defined as the safeguarding of public health, equitable access, and availability of water as a factor of production.

Yet by the early 1980s, efficiency began to be prioritized, formally, in water supply management. Efficiency in water production received greater emphasis in the period following the creation of the Regional Water Authorities in 1974, reflecting then-dominant trends in administrative reform, and growing concern over water quality rather than the previous focus on quantity (Hassan 1998; Jordan, Richardson et al. 1977; Penning-Rowsell and Parker 1983). Efficiency, during this period, was conventionally defined in engineering terms as operational efficiency—efficiency in water production.

Cost efficiency—the reduction of costs per customer or per unit of volume of water produced—was also increasingly important in the 1980s. After 1974, the rate-support grant, through which water bills were subsidized by the state, was removed. Faced with the costs of renewing deteriorating Victorian-era infrastructure, but operating under strict monetarist constraints, the government was left with two options: either devolve the costs of water quality improvement on to consumers through large increases in water bills; or continue to allow water quality to deteriorate. The government held to its decision to prioritize the PSBR over water quality improvements. Whilst allowing increases in water bills—sometimes higher than the inflation indicator Retail Price Index (RPI) but often lower than required to meet investment requirements—it encouraged greater cost efficiency in the water sector as a means of offsetting capital shortfalls (Simon 1986). The Water Act of 1983, for example, was predicated on the assumption that consumers were best served 'by an efficiently run operation, providing the requisite service at least cost' (MMC 1981: 264). New efficiency measures were introduced; for example, job shedding in the industry dates from the early 1980s with employment levels peaking in the early 1980s and reducing substantially before privatization in 1989. Well before privatization, cost–benefit analysis was introduced to the industry (Synnott 1985: 70).

With privatization, the prioritization of efficiency was formally integrated into the regulatory framework. Promoting efficiency is one of the primary duties of the economic regulator of the water industry (s. 2.3.d Water Industry Act 1991). The introduction of the concept of economic efficiency represented a radical change in the principles of water regulation; as one interviewee noted:

> Until privatization, most of the water companies didn't think economics had anything to do with them. The employees were mostly engineers who were unsympathetic to economics, because economics is about uncertainty, but engineering is about certainty, that 'the bridge must not fall down'.[1]

Operationally, the concept of economic efficiency occupies a central role in the regulatory framework, through the system of 'price-cap' regulation employed by the economic regulator. This method of regulation, designed by Treasury economist Stephen Littlechild in the mid-1980s, has been applied to all of the privatized British network utilities (Littlechild 1986, 1988). Unlike American-style rate-of-return regulation, in which dividends are capped, utilities' maximum price increases are capped in a system conventionally referred to as 'RPI $-$ X = K' (a simplified version of the formula used to determine the price increases). Some analysts have argued that the differences between rate-of-return and price cap regulation are not as significant as was originally expected—largely because the UK cost accounting that underlies the setting of price caps requires implicit assumptions to be made about the needed rate of return on new investments. Nonetheless, according to the architects of privatization, price-cap regulation was deemed to be 'preferable to rate of return regulation of profits because it is simpler, less expensive and interventionary [*sic*], and less vulnerable to 'cost-plus' disincentive effects' (Littlechild 1986: 2). Under price cap regulation, each company is required to cap or limit the annual rate of growth of the (weighted average) price of its regulated services by a factor K, which is calculated as the rate of growth of the Retail Prices Index (RPI), minus an efficiency factor (X); prices will normally rise slower than the rate of inflation.

The economic regulator of the water industry, Ofwat, uses a variant of this formula, where charges are controlled by the price limit formula RPI + K + U, where U is any price limit not taken up in previous years, and K incorporates the cost of expenditure on water quality improvements minus an efficiency

---

[1] Former water industry manager, interview with author, September 1997.

factor: K = Q − X. During the first two reviews of the industry by the regulator, | Q | > | X | due to high capital expenditure required to redress past under-investment, and to meet increasingly stringent EU water quality legislation. In the case of the water industry, RPI + Q − X > 0, with prices rising above the rate of inflation, unlike other privatized utilities.

The price cap system operates on the assumption that the regulator, gathering information about firms' performance and required investment, can set an upper limit on price increases that allows an efficient company to achieve a reasonable revenue stream (Glynn 1992). The merit of the system arises from the way in which this 'capping' of prices encourages efficiency. Price caps are calculated by the regulator every five years in the 'Periodic Review' process, and set in advance. In order to calculate the price caps, the regulator employs econometric models and detailed assessments of individual company performance to identify potential reductions in operating, capital maintenance, and capital enhancement expenditure (Ofwat 1998c). Thus, in theory, the incentive for a water company to increase efficiency arises from the fact that companies can increase profit by increasing efficiency, thereby retaining expenditure in addition to the revenue allowed by their price cap. Comparative competition thus relies on a set of benchmarks, which are in theory a function of all firms' performance; any one firm's allowed price increase is thus not solely a function of its own actions, thus diminishing (if not eliminating) the scope for strategic behaviour on the part of the firm.

Operationalizing valuation techniques

Operationalizing price cap regulation requires the implementation of mechanisms which allow for efficiency to be measured and maximized. If optimizing water use is defined as maximizing efficiency in water use, then optimal water management requires a price mechanism that accurately reflects the value of water and the costs of water production. Two sets of techniques are required: techniques that will enable the correct valuation of water by taking full account of the costs of water's production and consumption (environmental economic accounting), internalizing environmental costs and benefits (Pearce 1976; Garrod and Willis 1994; Winpenny 1994); and techniques that will enable the correct pricing of water by employing a price function which captures costs (Zarnikau 1994; Scott 1995; Hall 1996) and establishing a mechanism by which price signals can be communicated to consumer and supplier (metering) (Russac, Rushton *et al.* 1991).

Pricing and metering water

In addition to properly valuing water in order to establish its true costs, commercialization requires the correct pricing of water, which in turn entails the establishment of a price function and a mechanism for communicating price signals to consumers. In order to signal cost-reflective prices accurately to consumers, some form of volumetric metering is required (Herrington 1996b). Accordingly, 100% penetration of meters into domestic properties was envisaged in 1989; full metering was originally required of the water supply industry by the year 2000. Meters have been progressively installed in domestic properties since 1989, but the rate of meter installation was much slower than expected (this process is detailed in Chapter 6), and met with resistance on the part of consumer advocacy groups, some politicians, and nearly all water companies, with the exception of Anglian Water—in one of the tightest resource positions of all of the companies—which began mentioning metering policies as part of its water conservation strategy in its corporate briefings as early as 1993. Anglian was, however, the exception. By the late 1990s, the government had quietly dropped the obligation on the part of water companies to meter all customers. By 2001 domestic metering penetration levels had not yet reached 20%.

Understanding the difficulties with which the obligatory metering requirement was faced requires reference to the communicative function of water meters. Meters are a means of introducing measuring points into every home, opening up the 'black box' of consumer demand to quantification; accordingly, 'the *set* of meters is a powerful instrument of control' (Akrich 1992: 217). This move is understandably of great interest to companies seeking to manipulate demand so as to minimize operating expenditure, as indicated by the growth over the 1990s in the number of surveys monitoring domestic consumption (Turton 1995). Yet a meter is a two-way means of communication; it also provides a means whereby control is destabilized. Currently, water utilities are ensured of stability in future revenue streams given their captive monopoly domestic markets and property-value based charging systems. An increase in metering, in short, may imply a decrease in the predictability of companies' revenue streams.

Metering may also affect customer behaviour in unexpected ways. The economic regulator supports metering, arguing that it provides customers with a choice, insofar as a meter provides them with information enabling them to control bill levels. Simultaneously, meters provide a means whereby price signals can be sent to consumers; meters act as a reciprocal medium of communication. The environmental and economic regulators argue that meters combined with cost-reflective pricing will encourage conservation. However, a decrease in consumer responses to appeals to conserve water during times of drought, as observed during the Yorkshire drought of 1995, when demand actually increased after company appeals to reduce usage, may be an unexpected result of metering in some cases. Although few studies exist, some water industry managers doubt that metering will act uniformly to reduce consumption, and mention the possibility of metering 'bounce back' (a rise in consumption associated with changing behaviours stemming from metering justifying consumption as long as the water bill is paid). Unpredictability of revenues and customer behaviour is the Janus face of the increased surveillance opportunities open to companies through the installation of domestic meters. Companies' resistance to metering is partly explained by this surveillance–security trade-off.

Valuing water as a scarce resource

Environmental economic valuation may be defined as the assignment of a monetary value to the environmental functions of a good or service that are not accounted for in the price mechanism. The justification for the introduction of environmental economic techniques stems from their putative ability to counter the 'perverse' economic effects of the non-market pricing of water. From an economic perspective, these effects impact negatively on the water sector (e.g. by delaying investment), the environment (e.g. through increased externalities), and the entire economy (e.g. by distorting growth). Environmental economic valuation will, it is argued, redress these perverse economic effects by more closely aligning the price of a good or service with its true value.

Valuation techniques generally entail what Grove-White refers to as 'surrogate valuation' (Grove-White 1997); techniques designed to simulate the market, through eliciting money values for environmental goods. One such set of techniques—contingent valuation—has become widely used in environmental economic assessments in the UK, through the use of methodologies such as 'willingness-to-pay' (for example how much is a clean river worth to you?), or 'willingness-to-accept' (what would you accept as compensation for that river being polluted?). The Environment Agency's approach to environmental valuation is a case in point:

the philosophy underlying [our] approach is that the value of a unit of water should be *maximised in its use*, whether this is as a drinking water supply source, for irrigation or for protection of the aquatic ecology. By incorporating valuations of the environmental and other social implications associated with changes in water use, the most *efficient* allocation of water and other resources amongst competing requirements should become more clear. (EA 1998*b*: 2; emphasis added)

The EA methodology draws on the results of existing environmental and social impact valuation studies using 'benefit transfer'; for example, assigning values determined through WTP surveys for one river, to a similar river in another catchment (RPA 1998). Having identified low, medium, and high transfer values, to be used either individually or together as part of a sensitivity analysis, all of the costs/benefits are identified for each water management option. The costs and benefits for each option are then summed, discounted, and incorporated into a 'resources planning guideline' together with capital and revenue or variable costs. The final figure represents the 'net present value' for an option; for example, in a recent EA assessment, a net present value of £13.6m. was assigned to a small chalk stream in the upper Thames river basin

(McPherson 1998). The figure was generated to support the EA's claim that abstraction from the upper reaches of the River Kennet, by Thames Water Utilities Ltd. (TWUL), should be reduced. As detailed in the following section, this particular figure merits further examination, as it was part of one of the most controversial investigations into environmental valuation in the water sector post-privatization.

The Axford case

The case of the debate over abstraction of water from the Axford area on the Kennet river is emblematic of that of many 'low flow' rivers in England and Wales (NRA 1993c). The Kennet is a tributary of the Thames, a small, very high quality chalk stream, whose quality and flow have decreased noticeably over the past few decades (McPherson 1998). Running through a predominantly rural, agricultural area, it has been controlled and managed by humans for the past 2,000 years (McPherson 1998), and supports a variety of human-managed environments—fisheries, water meadows, bird sanctuaries. TWUL, on the basis of a licence granted to its local authority predecessor in the 1960s, abstracts 20.5 Ml/d from a borehole near the river. The majority of the water is piped out of the Kennet's catchment to supply the nearby, rapidly growing town of Swindon, where demand exceeds the water resources within the locality.

In attempting to force TWUL to reduce its abstraction in the Axford area, the EA rejected the company's application to renew its abstraction licence, and required a reduction in the volume abstracted. On appeal to the Secretary of State for the Environment, TWUL noted the cost of finding additional sources: £6.2m. In its evidence to the DOE Inspector during the ensuing review, the EA employed benefit transfer and CBA techniques to derive its 'net present value' figure for the river, composed of £0.4m. in 'use values' (related to the direct or indirect use of the resource), and £13.2m. in 'non use-values' (related to the conservation, preservation, and 'existence value' of the resource). The total NPV of £13.6m. far outweighed the costs to TWUL of finding new sources of water for Swindon, the EA argued; the benefits to be gained by reducing the abstraction far outweighed the costs.

In his decision, however, the DOE Inspector explicitly challenged the EA's environmental economic methodologies and reasoning (McPherson 1998). Although he accepted that 'there has been a significant fall in the general water levels in the River Kennet' associated with the 'fall in the associated water table', he disputed the correlation between this fall in water levels and the decrease in water quality. Other human activity—for example dredging, the construction of weirs, increase in use of nitrate-based fertilizers in the surrounding agricultural land—meant that abstraction could not be viewed as the primary cause of a decline in water quality. Although there was 'sufficient evidence of ecological harm to require the [minimum] flow constraint above the present figure' (McPherson 1998: 134), particularly given the impacts of increased abstraction in periods of low flow, such as summer, the abstraction itself need not be decreased. The reasons given were straightforward. First, the need for water:

Bearing in mind present leakage rates and the difficulty of reducing them, I have some doubts about TWUL's ability to achieve their target leakage . . . and to manage demand . . . at any rate, with the projected population increases [in Swindon] I can see no realistic prospect of reducing the present level of demand for water in the Swindon Sub-region. (McPherson 1998: 135)

Thus, although the inspector admitted 'the additional water is available' to replace that derived from the Axford abstraction, the abstraction, it was decided, should continue because 'the infrastructure necessary to carry the [replacement] water is not [available]' (McPherson 1998: 136), and the water from the current Axford abstraction was needed for Swindon. Critically, the cost of this infrastructure would be more than the recalculated 'value' of the River Kennet, according to the inspector's reworking and scaling down of the EA's calculations.

In focusing on the relative size and reliability of cost figures for different environmental options, the inquiry side-stepped the broader environmental policy objectives which the EA pursues (EA 1998*c*; 1998*d*), and which motivated its original decision to decrease the Axford abstraction. As had been noted earlier in the inquiry, the Axford abstraction supplies 17% of Swindon's demand; leakage in the Swindon area was calculated by TWUL to be approximately 18% of demand. The remainder of the (high quality) water abstracted from Axford is blended with high-nitrate concentration (low quality) water from another source, heavily contaminated with nitrates from agricultural run-off (Conrad 1990; Burt and Haycock 1992; Burt, Heathwaite, and Trudgill 1993), in order to produce water of potable quality. In the absence of source control to mitigate nitrate levels (Newson 1992), and leakage reduction in Swindon to reduce demand, the Axford abstraction represents a cheap source of high quality water for TWUL. From the perspective of the EA's policy on sustainable development, this makes no environmental sense; from the environmental economic perspective, however, the decision is (if not reasonable) entirely rational.

Critiques of surrogate valuation techniques

The Axford case demonstrates the perils of playing the 'numbers game'; the EA gambled—perhaps with a flawed methodology, as some commentators have argued (Bateman, Longford *et al.* 2000)—and lost. The application of environmental economic techniques entails the assumption that in order to make a 'rational, non-arbitrary choice between options there must exist some common unit of measurement through which the relative importance of goods can be ascertained' (O'Neill 1997: 77). The relative 'brittleness' of surrogate valuation techniques (Grove-White 1997), and their exclusion of intangible, non-quantifiable evidence can, however, undermine these strategies. Surrogate valuation techniques perform a critical task of discursive boundary work, establishing what is to be enclosed within definitions of the 'environment', open to valuation and thus eligible for protection, and what is not to be valued, the realm of intangible nature, still subject to territorial effects. Our techniques for measuring the world, in other words, shape how we perceive and define nature, by implying a choice as to which elements of nature are considered to be the 'environment' and worth valuation. Yet these definitions are open to challenge, as demonstrated in the DOE inspector's revaluation of the river in the Axford case.

The Axford case also reveals the contradictions arising from the disjuncture between environmental economics and environmental policy, when market techniques are applied to non-marketized spheres of existence. The (often unarticulated) goal of surrogate valuation techniques is the maximization of total preference satisfaction (benefit over cost) of all individuals (Jacobs 1997). These preferences are expressed as private choices, and are treated as commensurable with other choices made through the market. In this sense, valuation techniques are 'value-articulating institutions'—'arenas or mechanisms through which people can articulate the different values they place on different goods' (Jacobs 1997: 213). The difficulty with environmental valuation lies in its assumption that a model widely used for private goods is also appropriate to public goods. The underlying political theory is, of course, derived from neo-classical economics: autonomous, welfare-maximizing individuals whose preferences are ethically incontestable, and should be met by public institutions. Based on these premises, environmental economics seeks to address environmental problems through the commodification of the environment based on the premises of: methodological individualism; the rational economic utility-maximizing person; free, competitive markets; and exogenous preferences and technology (Jacobs 1994).

However, advocates of environmental economic methodologies often overlook the fact that environmental goods are public goods in three senses: collectively consumed and indivisible; an object of ethical concern (thus requiring a public arena for debate); and a common good which has an 'existence value', whether or not people actually make use of it (Jacobs 1997). Surrogate valuation ignores the ways in which other factors, such as lack of information, time constraints, institutions, and culture affect behaviour. It also fails to acknowledge that individual practices encompass more than economic activity, or the fact that economic practices do not necessarily take place in the market (Gibson-Graham 1996); indeed it assumes the contrary. Economists who pose similar questions concerning its underlying assumptions have made critiques of surrogate valuation (Aldred 1994; Bowers 1993; Jacobs 1994). Environmental philosophers' critique of surrogate valuation rests on their distinction between values as norms, rather than utilitarian concepts (O'Neill 1997). These critiques have been translated into queries regarding methodological consistency (Bilsborough 1997; Burgess, Harrison, and Clark, 1998). The inconsistencies embedded in environmental economic conceptions of human behaviour and values of are revealed, for example, in disparities between 'willingness-to-pay' and 'willingness-to-accept' compensation values in surveys, reflective of the lack of clarity regarding 'the circumstances under which cost-benefit analysis can be legitimately used to set environmental objectives, targets or prices to generate environmental output' (Bilsborough 1997: 89).

Leaving aside, for the moment, questions of methodological consistency, and, more generally, of the validity of assigning economic *values* to individual *preferences* (Sagoff 1988), the distinction between market forces, and market-simulating techniques bears further scrutiny. From the neo-classical resource economics perspective, a market is a social institution, one which 'makes available to affected parties the opportunity to negotiate courses of action' (Dasgupta 1990). A market is reciprocal; both buyers and sellers send signals to one another. Surrogate valuation, on the other hand, is not reciprocal; it merely attempts to simulate the behaviour of individuals as if a market existed. The basic assumptions of the neo-classical world (resources are divisible, substitutable, and can be privately owned; environmental-economic systems are characterized by continuous evolution between stable equilibrium; system changes are continuous and perfectly reversible) do not hold true for the majority of cases in which surrogate valuation techniques are applied (Norgaard 1985). Proponents of these techniques fail to distinguish between market situations, and market-simulating techniques, conflating the two in the application of environmental economic logic to decisions about resource management. In the absence of a free market, however, the application of market-simulating techniques may have unforeseen effects.

These critiques aside, environmental economic valuation is increasingly central to water regulation in England and Wales. An important application of environmental economic valuation is as a means of broadening the cost-benefit exercises undertaken to assess the various options to meet predicted water supply shortfalls (DETR 1998*b*). Cost–benefit analysis, in the environmental economic lexicon, has been extended to include 'intangible' values, incorporating, for example, people's aesthetic appreciation of nature, or valuing of the existence (rather than access to or use of) certain species or landscapes.

The adoption of environmental economic methodologies by the environmental regulator is in part a political tactic. As one interviewee noted:

The EA were very sore about the last [Periodic] Review [of water prices and investment plans in 1994]. It was not well handled by the EA. The environmental obligations, for example, were overpriced, and so Ofwat took a look at the costs, and then left them [i.e. the environmental obligations] out. The EA are being very aggressive this time around [in using environmental economic methodologies].[2]

[2] Environment Agency, interview with author, July 1997.

This view accords with Grove-White's description of the rationale behind the growing use of environmental valuation by policy-makers (Grove-White 1997). In Grove-White's assessment, environmental valuation techniques have increased in popularity due to their ability to function as a tactical tool for policy-makers. In particular, environmental valuation allows previously marginalized environmental policy-makers a tool with which to pursue greater legitimacy; 'when the DOE brought in environmental economics in the late 1980s, it was a sign to the Treasury of "it's our numbers against yours now" ' (Grove-White 1997: 23). As Grove-White notes, environmental NGOs such as the RSPB and CPRE, and statutory agencies such as English Nature, the Countryside Commission, and the NRA 'have all tended to explore the tactical possibilities of the use of surrogate valuation methods for advancing their aims' (Grove-White 1997: 28). For the Environment Agency, as a lesser partner in the post-privatization regulatory framework, environmental valuation provided a means with which to 'fight numbers with numbers', inserting the environment into Ofwat's cost-benefit calculations and investment plan decisions.

This tendency has been reinforced by a trend towards greater 'marketization' of public sector activities engendering a new 'audit' mentality in government activities (Du Gay and Salaman 1992; Power 1994). Within government, the effect has been to enhance the role of the Treasury as arbiter of public priorities. This privileging of economic accounting of environmental issues has been codified in the water regulatory framework, as the Environment Agency has a duty under Section 39 of the Environment Act (1995) to 'take into account the likely costs and benefits' of any requirements it imposes, which it interprets as meaning *environmental* costs and benefits (EA 1998a). Ofwat, on the other hand, has no statutory obligation to promote sustainability or consider environmental externalities when deciding upon company's investment plans. Rather, equity consists of achieving an appropriate balance between the interests of shareholders, consumers, and water companies. Environmental obligations which impose too great a cost burden on consumers are resisted by Ofwat, which treats WTP surveys and other EA studies of consumers' attitudes towards the environment with 'great caution' (Ofwat 1998b: 10). Accordingly, water companies' requests for approval for investment to meet environmental obligations, rather than statutory water quality obligations, were for the most part not approved by Ofwat at the first Periodic Review.

Pricing and metering: from planning for growth to managing scarcity

Until recently, water resource planning was characterized by supply-led solutions. The typical planning scenario was to 'commission a comprehensive study of resources; project the demand on an 'unconstrained' scenario; consider the various supply-augmentation options; recommend that which meets project demand at the least cost; and implement the scheme through public agencies, and at a subsidised price' (Winpenny 1994: 22). During the 1960s and 1970s, crude, linear extrapolations of short-term demand curves provided the basis for demand projections, legitimating large-scale resource development (WRB 1974). The priority of water resources management was to maximize security, albeit with reference to costs. Water resources management was predicated on the implicit assumption of the abundance of water; water conservation was systematically prioritized only during droughts.

This set of assumptions has begun to be undermined in the past two decades, during which there has been an exaggeration in the north-west to south-east rainfall gradient across the British Isles; a more distinct partitioning of annual rainfall totals between the winter and summer half-years; and exceptionally mild weather encouraging 'exceptionally high rates of evaporation' (Marsh 1996). These tendencies, which show a broad consistency with a number of climate-change scenarios (Arnell, Jenkins, and George 1994; DOE 1996*b*), have raised questions regarding the resilience of water resource management strategies (Marsh 1996).

Even neglecting the possible effects of global climate change (Arnell, Jenkins, and George 1994; Herrington 1996*a*), long-term shortages had begun to be predicted by the early 1990s (NRA 1992*a*; 1992*b*; Cryer 1995). By the end of the decade, the Environment Agency was predicting water deficits for five of the ten existing water authority regions in England and Wales within twenty years (EA 1998*d*; 1998*e*). These regions lie in the south and east of England, areas of highest demand yet lowest supply of fresh water (EA 1998*e*); shortages would arise due to increased demand due to population growth and increasing demand per capita, pollution of water resources, and increasingly stringent water quality regulations which effectively reduce the amount of potable water available for use. These shortages would be compounded by ageing infrastructure (Maclean 1991; Cowan 1993), and perhaps by increasingly volatile climatic patterns. Indeed, the decade following the privatization of the water industry coincided with a period of volatile climatic patterns, of which 1995, containing the driest 5-month period on record following the wettest 30-month sequence in the entire British rainfall series (which extends back to 1869), was the most dramatic (Marsh 1995). Recent extremes in climatological and hydrological conditions, including more numerous and severe dry periods, have focused attention on the possibility of both short- and long-term water scarcity.

In response, 'demand-side' water-management mechanisms are increasingly being advocated and in contrast to the near-exclusive reliance on supply-side strategies in the past. The NRA (predecessor to the EA) developed extensive demand management strategies in the mid-1990s (Demand Management Centre 1994; NRA 1992*a*; 1992*b*; 1993*a*; 1993*b*; 1994*b*; 1995*a*).

Options for meeting the predicted supply shortfall are being considered on both the 'supply' or the 'demand' side, attempting to increase the water available for use and/or attempting to decrease overall demand (OXERA 1997). Companies are 'required to demonstrate that their plans incorporate the least-cost solution to balancing supply and demand' (Ofwat 1997*d*: 3) as set out in the Ofwat publications *Paying for Growth* (Ofwat 1993*a*). On the supply side, companies and regulators are considering both conventional resource developments, such as reservoirs, and more unconventional ones, such as desalination and inter-basin water transfers. On the demand side, both internal (for example, leakage reduction) and external (for example, installation of water-efficient appliances in domestic properties) demand management measures are being evaluated (Hunter, Rowe *et al.* 1996). The most appropriate mix of these measures is being evaluated within a single economic framework, which includes, for the first time, the environmental costs of resource development (EA 1998*b*). Proposed changes to the abstraction licensing regime which would set prices for water abstractions determined via environmental economic techniques would extend valuation techniques further into the resources management framework (see, for example, DETR 2000*b*).

In sharp contrast to past practice, the regulatory framework is increasingly oriented towards planning for scarcity, rather than planning for growth. Yet it should be emphasized that this scarcity of resources is simultaneously natural (resulting from changing climatic patterns, for example) and social (due to, for example, the increasing expense of water abstractions from environmentally sensitive wetlands). Environmental economics is predicated upon the assumption that resources are scarce, but the introduction of environmental economic techniques also contributes to the social creation of scarcity through enabling valuation of resources in monetary terms. In doing so, environmental valuation techniques serve to naturalize water supply scarcity as a social phenomenon.

This transition from a recognition of water as an abundant to a scarce resource has recently been enshrined in UK legislation. The modifications made in 1999 to the Water Industry Act (1991) altered the bases upon which consumer-charging mechanisms operate. Under the revised Act, consumers may choose to be charged by meter, or to opt out of metered supply, but may exercise this choice only if water in the home is not used for non-subsistence purposes (such as garden-watering, or a pool), or if the property is in an 'Area of Water Scarcity', as designated by the Secretary of State for the Environment following application by a water company (S.I. 1999 No. 3442). The legal definition of water scarcity is relative to available resources, 'likely' demand, and the measures a company could undertake to meet demand. In these areas, water companies will be able to undertake measures not normally permitted under the regulatory framework, such as compulsory metering and, in some cases, cost-reflective pricing.

Long run marginal (LRMC) cost pricing

The rationale for cost-reflective pricing lies in its promise of optimizing water use, via the technique of marginal cost pricing. Marginal cost can be defined as the cost resulting from a unit increase in production at the margin. Marginal cost pricing entails setting the price of a good equal to the marginal cost of production. Where marginal costs and benefits are equal, according to neoclassical theory, efficiency is maximized (Ofwat 1997c). For example, in an area where there is no margin of supply over demand, and any new demand will require new resources which will be very expensive to obtain, the marginal cost for new demand will be very high. Pricing water at marginal cost rather than average cost would encourage customers to conserve. According to its proponents, valuing water appropriately should lead to demand management in the absence of prescriptive state regulation; environmentally sustainable solutions are the promised outcome.

The biophysical characteristics of water, however, pose challenges to implementation of this logic. Costs vary temporally (for example due to peak loading), spatially (for example due to distance from the source), and seasonally (for example due to raw water availability); scaling water charges to the appropriate spatial and temporal resolution thus poses significant practical difficulties. Given the lack of metering, higher demand does not generate additional revenue; resource development must be funded out of revenue from non-volumetric related charges, the cost of which can not be proportionally assigned according to level of water usage. In addition, water resource investment is 'lumpy' rather than smooth; accordingly, rather than the smooth marginal cost function of the economist's ideal world, the marginal cost (MC) function is 'spiky'. Accordingly, smoothing out this MC curve using 'long-run marginal-cost' analysis, has been Ofwat's preferred methodology for determining an optimal solution to the supply-demand balance.

Long-run marginal cost attempts to smooth these spikes in the price curve (reflecting the 'lumpiness' of water investment) by estimating the marginal cost over a long timescale (for example 10 years), using the cost per unit of capacity of the next resource development (or treatment plant) to be built (Hall 1996; Scott 1995). LRMC is the methodology that Ofwat specifies WaSCs and WoCs must use in determining the 'solution' to any gap between supply and demand. That is, the most optimal economic solution, as determined by LRMC, must be followed in determining what mix of water management options (e.g. building a new reservoir, reducing leakage, encouraging xerophytic gardening) companies will use in making up any water supply shortfalls. When used to set tariffs for water supply, LRMC implies that 'tariffs should be based on the marginal cost of supply, interpreted as the cost of adjusting long-term capacity caused by a given changed in demand' (Winpenny 1994: 9).

LRMC pricing has been implemented by some water companies in England and Wales. For example, one WaSC in the North of England, with excess capacity and thus a low marginal cost, has reduced its volumetric charge, while increasing its standing charge—effectively encouraging consumers to use more water (ESRI 1994). Conversely, one WaSC in the South of England, with a high LRMC, is lowering the standing charge and raising the volumetric charge, sending the opposite signals to consumers in an area where supply barely meets demand (ESRI 1994; Scott 1995). As these examples demonstrate, it is not always economically efficient to conserve water. Although the Environment Agency and environmental groups such as the Council for the Protection of Rural England support metering and other demand-management innovations, they recognize that 'once customers are metered the incentive for companies to restrict demand growth vanishes; with controlled unit prices companies [will] make more money from selling rather than saving water' (Rees and Williams, 1993: 23). In cases where there is a surplus resource, it may be more economically efficient to encourage increases in use by customers, and to keep encouraging increases in demand until all excess capacity is used up.

Demand-side management

Nor is it necessarily economically efficient to reduce leakage. 'Demand-side management' (DSM) techniques, such as leakage reduction, and consumer conservation campaigns, are, from an economic perspective, techniques for smoothing the discontinuous cost curve. DSM should be applied in situations of deficit or near-deficit of water, and only when the marginal cost of the next unit of water is thus very high (London Economics 1997). In other words, DSM is useful to water companies as a short-term response to critical shortages, and as a means of 'peak-lopping', smoothing out the cost curve before a new resource is brought on line, or for containing demand in a water-'stressed' zone or locality instead of engaging in expensive new mains renewal or refurbishment. For companies that have a surplus of water resources, and thus a low marginal cost, it is cheaper to treat and leak water than to incur operating expenses fixing leaks. Yorkshire Water, for example, in its first study of an 'economic level of leakage', indicated that leakage rates of 27% were optimal.[3] As one senior manager from another water company noted, having been asked about company leakage rates reported at nearly 40%:

Leakage of treated water is not a problem cost-wise. After all the money spent on pipes, a bit of money spent on chemicals etc. is incidental. The water doesn't cost that much to treat. So companies don't really lose a lot of money this way, until resources are tight.

[3] Yorkshire Water reported a figure of 363 Ml/d as its Economic Level of Leakage in 1999/2000 (Yorkshire Water 1997). Based on the amount of water supplied by Yorkshire in 1996/7, 1,351 Ml/d (WSA 1998), this represents a leakage rate of 27% of water put into supply. It should be noted that the water supply industry prefers not to measure leakage in percentage terms, given that this masks important variables; for example, percentage figures to not take into account length of pipe per customer served and are therefore biased against companies serving large rural populations.

We are having to decrease leakage because of political reasons and public perception. It was never considered a strain on resources. It is cheaper to go on treating and leaking as long as water is plentiful.[4]

DSM, in other words, is not always economically efficient for companies. As DSM is mostly used on the metered side of the business, it has direct cost implications for companies who are charging rates above marginal cost. Companies have therefore been, on the whole, reluctant to introduce widespread metering for domestic consumers, particularly in regions where marginal costs are low and metering would reduce bills, implying a drop in revenues. They have also been reluctant to introduce marginal cost pricing where marginal cost is less than average cost, because this would imply a drop in revenues, given that LRMC satisfies the allocative function of price but does not ensure cost recovery for which additional charges must be imposed (Scott 1995; *Utilities Journal* 2001c).

This observation, it should be noted, directly contradicts analyses of water industry trends (Guy and Marvin 1996a; Guy and Marvin 1996b) that highlight the shared interest of infrastructure providers and users in managing demand and 'more environmentally sensitive forms of network management' (Guy and Marvin 1996b: 137). Depending on the marginal cost of water in a given region, operational efficiency ('conservation'), and economic efficiency are not necessarily simultaneously maximized. The implications—for security of supply, operational efficiency, and drought management—are explored in detail in Chapter 5.

The justification for this prioritization of efficiency is its supposed effect on consumer welfare—defined in economic terms. A minimization of prices for a given level of service is predicted to be the result of efficiency gains in water services provision. In legislation and in practice, it should be noted, the minimization of prices is not an explicit goal; the promotion of efficiency is an explicit goal (and a primary duty in the case of the economic regulator), from which the minimization of prices for a given level of output is expected to result. The original regulatory framework post-privatization contained no explicit mechanism for addressing the question of the *acceptable* level of these prices, in contrast to the previous pricing regime.

Recent guidance from Ofwat recommending that companies submit LRMC estimates for each supply zone (where these are reasonably geographically independent) (*Utilities Journal* 2001c) raised concerns on the part of consumer advocates, noting that differentiation of charges within regions might lead to significant price rises for rural consumers. The current consensus—that universal metering is theoretically desirable but impractical and expensive—implies that temporal and spatial cross subsidies will continue in the water sector. The emphasis on economic equity has shifted the balance of these

---

[4] *Financial Times* (1996b) 'Water competition plan axed: Moves for national grid and wider choice hit technical problems', 30 Dec. 1996, 14.

cross-subsidies. Increasingly stringent regulation, and the direct intervention of the government on numerous occasions in the late 1990s have sought to mitigate some of the most politically unacceptable effects—such as an increase in 'water poverty'—as will be explored in detail in Chapter 6.

Liberalization: competition proves elusive

Price-cap regulation does not necessarily require the liberalization of the water industry in order to function, as 'comparative' or simulated competition may be employed. Comparative competition entails the calculation of potential efficiency gains not only through reference to individual companies, but also through the relative ranking of company performance. The setting of price caps is thus dependent upon simulated or 'comparative' competition. Here, efficiency targets serve as a 'proxy for a competitive market' (Ofwat 1998d: 49). With price caps set in advance, competition amongst companies occurs relative to some efficiency 'yardstick' calculated by the regulator, backed up by the threat of takeover in case of poor performance, with scorekeeping by the capital markets and occasional refereeing by the regulator. The profit motive is, in theory, harnessed by comparative competition-driven price cap regulation to drive efficiency gains and reduce costs. Periodic Reviews are the mechanism by which these cost reductions can be shared with consumers in the form of lower prices and improved levels of service. Competition is assumed to be a better mechanism of ensuring efficiency gains than command-and-control regulation, legislation, or moral suasion.

However, comparative competition is viewed by many within the industry as a 'pale and sickly relative of market competition' (Summerton 2001: 23). Competition in production and retail is viewed as the best means of promoting efficiency and protecting consumers against the abuse of monopoly power. The economic regulator has repeatedly asserted the view that customers 'shouldn't have to rely exclusively on comparative competition' (Garrett 2002: 6). The commitment to increasing competition is reflected in the statutory duties of the economic regulator to facilitate competition.[5] Five types of competition can be applied to the water industry: surrogate competition (for example comparative); procurement competition (for inputs such as capital); competition for corporate control (by mergers and takeovers); competition *for* the market (for example franchising); and competition *in* the market—i.e. product market competition (Cowan 1997a; Vass 2002). Product market competition may occur via common carriage (competitors using incumbents' assets for service delivery); cross-border supply (customers in one licensed area connecting to a network from another); and inset appointments (see below) (Ofwat 2002b).

[5] Water company manager, interview with author, August 1997.

These different types of competition have not been all or equally applied in the water industry in England and Wales. Competition for the market has been limited to date given the exclusive licences held by water companies on 25-year terms. In the first few years following privatization, comparative competition through the periodic review process, and competition for corporate control through the capital market were the dominant forms of competition. In this phase, mergers and acquisitions resulted in a concentration of the industry and a reduction in the number of companies—particularly the smaller water-only companies. All water companies are now large enough that any proposed mergers would result in an automatic referral to the Competition Commission. In the last three 'water-to-water' mergers cases examined by the Monopolies and Mergers Commission (now the Competition Commission), the MMC concluded that 'water-to-water' mergers would result in substantial detriment to the comparative competition regime. The MMC prohibited the mergers citing the need to retain a sufficient number of comparators in order for the economic regulator to carry out comparative competition. The economic regulator has indicated that he would be reluctant to see any further reduction in the number of comparators (down to 23 from the original 39 at the time of privatization) (see, for example, Fletcher 2001*b*; 2002). The requirement for a minimum number of comparators thus appears to weaken competition for corporate control.

Here, the regulator faces a dilemma at the heart of price-cap regulation: on the one hand, the preservation of a sufficient number of distinct water suppliers is necessary to underpin comparative competition; on the other hand, the need to retain the threat of takeover as a 'spur to efficiency'. Competition in the capital market is an essential complement to yardstick regulation, because 'the spur to efficiency is sharpened by competition in the capital market, including the threat of takeover' for badly performing companies (Littlechild 1986: 3). Takeover, however, reduces the number of comparators available to the economic regulator for use in his comparative competition model. Mergers in the water industry are therefore tightly controlled, and are automatically referred to the Competition Commission when above a certain size threshold. Ofwat's need for a sufficient number of comparators is a consideration in the Commission's decisions. Indeed, at the time of writing, the takeover of Southern Water by Vivendi Environnement (already well represented in the UK water sector) was rejected by the government (Fletcher 2003). In DETR's consultation on competition, the Department noted that the government was not proposing to amend the existing merger provisions, but to keep them under review and to allow/encourage competition to develop elsewhere in the industry ((DETR 2000*c*).

Other forms of competition have accordingly been the focus of the water supply industry. Procurement competition is important. The use of out-sourcing for activities ranging from pipe maintenance to information technology and customer call-centre management has been widespread in the industry.

However, proponents of product market competition assert that no other form of competition, including yardstick regulation, can increase efficiency, reduce prices, and stimulate innovation to the same degree (Helm and Jenkinson 1997; Robinson 1997; Cowan 1997a). Proponents of comparative competition maintain that capital markets are better assessors of water company performance than regulators. It is argued that 'City' scrutiny, backed up by share price movements as measures of performance, will ensure efficiency, effectively substituting competition in the performance of water managers for competition in the product market (Littlechild 1988). However, while City scrutiny and shareholder pressure encourages profit maximization, they do not necessarily encourage efficiency maximization. Profits in the water companies increased rapidly during the 1990s, and all water companies have paid dividends to their shareholders well above the average paid to stock-market investors (Ofwat 1996d). Sustained high profits and dividends levels have contributed to the perception that shareholder-driven incentives will trump consumer-driven incentives in the absence of product-market competition.

This is one of the reasons why government policy post-privatization shifted towards more explicit encouragement of product market competition, rather than viewing regulation as a permanent substitute for direct forms of competition (Vass 2002; DETR 2000a). Although product market competition has been introduced into most of the other privatized utilities, it remains extremely limited in the water supply industry. Introducing product-market competition into the water industry is hindered by technical factors, related to water's biophysical characteristics. Water is a 'heavy' product; hence the lack of an integrated national network. Water is a flow resource, required to fulfil environmental as well as public health functions; the differences in quality between different water supply zones pose particular problems for common carriage competition (Rowson 2000). The environmental and health and safety risks posed by water resources abstraction, supply, and disposal may also favour a unified management structure through the supply chain (Vass 2002). Regulatory complexity, with responsibility divided between regulators and various government departments, implies that no clear line of accountability exists for the sector overall, and that the transaction costs of regulating competition would be high. The structure of government, with water supply system ownership, management, and regulation varying between the constituent parts of the UK (England and Wales, Northern Ireland, and Scotland) is also perceived to pose greater barriers to some forms of competition—particularly common carriage (Vass 2002).

Additional difficulties in implementing competition are posed by the pricing regime. The present tariff regime includes high degree of effective cross subsidy, given that 80% of domestic customers are not metered and pay bills linked to the rateable property value of their home. Correct price signals are deemed necessary to enable competition between water companies to drive down costs, and to encourage customers to conserve where water is scarce.

Hence optimized price-cap regulation requires the correct valuation of water and metering, whereby price signals can be communicated to consumers. However, spatial disaggregation of pricing will lead to significant variations, given that regional costs are not reflective of LRMC; given the higher costs of supply in rural areas, spatial disaggregation may conflict with the regulator's duty to prevent discrimination against rural consumers. Given that competition requires subsidies to be 'unwound', the political imperative to retain prices at affordable levels and avoid large spatial variations perceived to be discriminatory against certain regions or classes of consumers is also a disincentive for competition.

Inset appointments

Committed to liberalization of the privatized utility sectors, the government devised legislation that permitted competition in the networked water product market at the time of privatization (see DOE 1992; 1995; 1996c). In the case of the water industry, this took the form of a provision for 'inset appointments' in the 1989 Water Act (later consolidated into the Water Industry Act 1991). Inset appointments are a mechanism by which the *de facto* monopoly of each water company can be challenged. Subject to approval by the economic regulator, inset appointments grant companies other than the licensed Appointee the right to provide water and/or sewerage services to a large user (a 'large user inset'), to a new building development with no connection to the existing water supply network ('greenfield inset'), or to a specific geographical area requiring a mutual agreement to change licence boundaries ('territorial inset') (Ofwat 1992b).

To date, inset appointments have mostly been limited to very large, usually industrial customers. Inset appointments are open both to water companies and to new entrants, and were expected to occur first in areas on the borders between the water companies, where a direct bulk water supply or sewerage connection would be easier to facilitate. However, this direct connection is not a necessary precursor to obtaining an inset appointment; 'virtual' competition is also permitted, as the existing Appointee continues to supply water and/or sewerage services, but is paid for doing so by the inset Appointee, who in turn bills the consumer. Competition through 'common carriage' via an expanding and integrating national network, rather than overlapping and separate company networks, was the envisaged outcome (DOE 1996c; Ofwat 1996g). Few inset appointments have, however, been issued to date; only eight had been issued with twelve more under consideration as of March 2001 (Fletcher 2001a).

One result of the introduction of competition via inset appointments was a strategic reduction in tariffs on the part of companies to large-scale users. Prices dropped for those cohorts of customers for whom competition is an option (large industrial users connected to the domestic water supply system) in order to ward off competition from inset appointments; the prices of domestic consumers (in the same tariff basket) have risen to cover the resulting loss in revenue (Consumer's Association 2000), implying an increased subsidization from domestic to industrial consumers. Although competition will provide large-scale water users with lower cost services, the domestic consumer is unlikely to be a beneficiary (Miller-Bakewell, 1998). In this instance, the regulator's imperative to encourage competition contradicts the requirement of prohibiting undue discrimination amongst classes of consumers. A further contradiction arises when the spatial scale at which competition would take place is considered. The unbundling of regional prices implies the reduction or elimination of subsidies to rural consumers; Ofwat's duty to promote competition in this instance conflicts with the duty to protect rural consumers, reflected in DEFRA's voicing of concerns about 'cherry-picking' of consumers should competition be widened in the industry.

Common carriage

Common carriage occurs when one service provider shares the use of another's assets, such as its pipe network or treatment works. In *Water: Increasing customer choice* (DOE 1996c), the government announced its intention to explore 'virtual markets' for large water users 'by allowing competing companies to monitor distribution over the single pipe system using electronic data interchange' (Guy, Graham *et al.* 1997: 201). This marked a retreat from the earlier commitment to introduce wholesale competition, as originally set out in the Water Industry Act (1991), and as extended by the Competition and Services (Utilities) Act (1992). The duty of the Director General of Water Services to develop competition in the provision of utility services and to minimize anti-competitive behaviour in the water industry was strengthened by the Competition Act 1998 (Parker 2000), which opened up the scope for common carriage by requiring companies to offer access to their facilities on reasonable terms or to provide objective justification in the case of refusal. Common carriage, it was anticipated, would be facilitated by the ability of companies to make formal complaints about anti-competitive behaviour by other companies under the Competition Act 1998.[6] Developers, for example, might lodge a complaint against a water company if it were felt to be abusing its dominant position when negotiating terms for new infrastructure works to new housing developments (*Utilities Journal* 2001*d*).

The development of common carriage to date has been limited in scale. The government retained a commitment throughout the 1990s to increasing competition. However, after prolonged consideration of the common carriage option, government announced restricted competition measures in March 2002. Only industrial and commercial users using large quantities of water

---

[6] By May of 2001, 44 complaints had been made to Ofwat under the provisions of the Competition Act 1998 (*Utilities Journal* 2001*d*).

(some 2,000 users) will be eligible for inset appointments under the new rules, much more restrictive than had earlier been anticipated (DEFRA 2002; Fletcher 2002).

The progressive retreat from full common carriage competition was motivated in part by safety concerns, given the difficulties posed by the problem of 'mixing' (most importantly the designation of responsibility for water quality incidents under common carriage), and also by recognition of the implications for pricing structure of the inherent conflict between goals of competition and efficiency (Vickers 1997). Industry critics argue that competition is resisted by the water industry as it 'would reduce profits and dividends and ultimately the directors' benefits' (Global Water Report 1997: 16). Water company managers, it seems, have resisted disturbances to the proverbial 'quiet life' of the monopolist. Officially, the government is continuing to consider the promotion of a more market-based approach to water resources and abstraction licensing, although falling short of tradable water abstraction permits, as had originally been proposed (DEFRA 2001*b*). Procurement competition is also likely to increase, given that out-sourcing of a variety of services and inputs is viewed as one of the best strategies for out-performing the regulator's forecasts. The economic regulator remains committed to introducing competition in the building of new distribution networks ('self-lay'), common carriage (albeit limited in scope), and abstraction licensing (DETR 2000*b*; Fletcher 2001, Ofwat 2001*a*). To date, however, competition in the water industry remains limited, particularly when compared with the degree of liberalization envisaged by the architects of the post-privatization regulatory framework (DETR 2000*a*; 2000*b*; Fletcher 2001).

Conclusions

Beginning in the 1980s, commercialization of water companies and water management practices has transformed the water supply industry and water management practices, entailing: the prioritization of efficiency as a goal of water management and the associated change in its definition; a shift from planning for growth to planning for scarcity; a redefinition of 'equity' in water pricing; the recognition of indirect consumption (for example, visual amenity) as well as direct consumption as having value which should be taken into account in the price of water; the formal recognition of the 'environment' as a legitimate user whose rights must be protected, balanced with those of the consumer and labour; the restructuring of employment relations, management structure, and internal company structure; and the reconfiguration of consumers not as citizens but as customers.

The post-privatization regulatory framework was premised on the assumptions that water is a scarce resource, that individuals seek to maximize utility, and that markets are the most efficient institutions through which to allocate resources. Recognizing that private companies seek to maximize profit, but not necessarily efficiency, the regulatory framework was designed not to mitigate but rather to harness the profit motive in order to incentivize efficiency gains, and to pass these gains on to consumers through lower prices and higher levels of service. In theory, this required substantial commercialization: widespread metering; product market as well as comparative and capital market competition; and pricing on the basis of economic equity with prices set as marginal cost. In practice, this has proved to be difficult; water supply networks—once built—are relatively easy to privatize, but much more difficult to commercialize. Water proved to be highly resistant to commercialization, not so much 'failing' the market as refusing to cooperate by throwing up trade-offs that have proved to be difficult to balance within the regulatory framework.

This chapter discussed several of these trade-offs in detail. As illustrated with respect to the Axford case, a contradiction may arise when employing techniques of water valuation between correct valuation of the environment and environmental conservation goals. When implementing demand-side management policies in the water industry, a (narrowly defined) concept of economic efficiency as employed in long-run marginal cost pricing formulas may in some instances counteract water conservation, in the sense of a sustained reduction of water demand. When introducing competition, regulators are faced with a trade-off between maintaining a sufficient number of comparators for comparative competition to function, and maintaining sufficient 'takeover' pressure on water company managers as an incentive for performance. In water pricing, regulators are confronted with a contradiction between the outcome of the application of the principle of economic equity (spatial differentiation of prices to facilitate competition), and a politically acceptable threshold of spatial homogeneity of bill levels. Implementing universal metering, a legal obligation on companies at the time of privatization, quietly dropped in the late 1990s, would have maximized allocative efficiency but not minimized the costs associated with metering installation and reduction in demand, with potential negative consequences metering for company revenues. In its policy and legislative decisions as to the balance of these trade-offs and contradictions, the government has retreated from the goal of the full commercialization of the industry. The reasons for doing so are detailed in subsequent chapters, which explore the contested politics of commercialization throughout the 1990s.

PART II

Re-Regulating the Water Supply Industry

PART II

Re-Regulating the Water Supply Industry

5

Privatizing Water, Producing Scarcity: The Yorkshire Drought of 1995

During the hot, sunny summer of 1995, reservoirs in England's West Yorkshire region ran dry. Despite restrictions on water consumption, demand soared to record levels. Yorkshire Water Services (YWS), the private company in charge of water supplies, obtained emergency permission to increase abstraction from local rivers. Amid concerns over deteriorating water quality and low flow levels, the environmental regulator of the industry carried out 'fish rescues' and artificially aerated some rivers (EA 1996d). Local schools, hospitals, and councils made contingency plans in the event of a failure in water supply. Local officials warned of health epidemics and fire hazards. Unessential uses, such as garden-watering and car-washing, were banned throughout the region. Standpipes (public water taps for access by multiple households) were erected, and YWS warned of 'rota cuts' (periodic cut-offs in domestic water supply). The region's reservoirs were empty, despite having been full only months before, after one of the wettest winters of the century.

Amid the ensuing public outcry, a large-scale reservoir refilling operation (using over 700 tanker trucks round-the-clock for five months) and a hastily instituted programme of infrastructure improvements were implemented. In spite of a massive publicity campaign to encourage consumers to reduce water use, regional demand for water soared, as did leakage rates from the mains system. The western half of Yorkshire, a region of three million people, was threatened with complete failure of its water supply system.

The end of the drought was officially declared in November 1996, with the lifting of all restrictions on domestic water use. YWS had narrowly avoided cutting off the water supply to several districts in West Yorkshire, but this did little to salvage its reputation as accusations of company mismanagement and underinvestment in water resources were borne out by two independent inquiries into the drought (Uff 1996; Ofwat 1996a). The chairman and managing director of YWS were replaced. The company was fined more than £40 million by the economic regulator of the water industry. However, the parent company's share price remained buoyant during and after the drought.

Despite the unexpected expenses imposed by the drought, YWS declared record profits in late 1995, and again in 1996.

The 1995 drought was the most extreme climate event, and the most negative public relations event, that had confronted the private water companies since privatization. As such, the drought raised questions about the implications of privatization for water resources and supply system management. The drought has been the subject of widely differing analyses. Meteorologists have investigated it as a climatic phenomenon (Marsh 1996). Political economists have examined the regulatory framework, the behaviour of the privatized company, and the economics of privatized utilities in arguing that the drought was human-made, an outcome of regulatory policies and company strategies that favour shareholders over consumers (Letza and Smallman 1996). Other observers have treated the drought as a discursively mediated crisis, in which contested and competing narratives of water companies and the media became partially decoupled from precipitation and sputtering taps to become symbolic of popular resentment of privatization of the water industry (Haughton 1998). Each of these approaches captures an element of the production of drought, at once 'real, like nature, narrated, like discourse, and collective, like society' (Latour 1993: 6). This chapter explores the interrelationships between the material, discursive, and social dimensions of the water supply system crisis in Yorkshire. In this chapter I unravel the relationships between these dimensions of drought through an analysis of the crisis in the water supply system in Yorkshire, an approach that seeks to demonstrate that water scarcity, in this instance, was socially produced[1]—through a complex admixture of meteorological conditions, regulatory decisions, and company and consumer actions.

The chapter begins with a summary of the simultaneous flows of water in Yorkshire and capital through the private water company during 1995 and 1996. In the following section I argue that the production of the supply system crisis was the outcome of three interrelated practices: meteorological modelling, demand forecasting, and corporate restructuring and the regulatory 'game' in the context of the regulatory framework which operated in the first quinquennium following privatization. In the final section, I briefly discuss the ways in which subsequent interpretations of the drought and its (mis)management, have figured in ongoing re-regulation of the water supply industry.

---

[1] This conceptualization of the production of scarcity draws on the theoretical discussion in Chapter 2, and also on research on the relationship between ideology and water management (Emel 1990; Emel and Brooks 1988; Emel and Roberts 1995; Emel, Roberts, and Sauri 1992), the production of nature (Castree 1995; 1997; Smith 1984; 1996), discourses of the environment (Castree and Braun 1998; Darier 1999; Downing and Bakker 2000; Hajer 1995; Litfin 1994; West and Smith 1996; Zimmerer 1993), and environmental regulation (Drummond and Marsden 1995; Gandy 1997; Gibbs 1996). For a more extended theoretical discussion, see Bakker (2000).

Flows of water and capital: a chronology of the drought

The drought of 1995 took regulators, consumers, and the private water company in charge of water supplies by surprise. The winter immediately preceding the drought (1994–5) was extremely wet; much of West Yorkshire, the region that would be worst hit by the drought, experienced rainfall well above the long-term average (LTA) in early 1995 (Uff 1996), and localized flooding occurred in some areas. By the end of the winter, groundwater levels were well above average, and reservoir stocks were close to 100%. In accordance with accepted practice, YWS began supplying demand through its regional network of trunk mains (the 'Grid') from upland reservoirs in West Yorkshire early in the summer. These gravity-fed sources in the Pennine hills represented a much cheaper source of water than that abstracted from lowland rivers, which incurred higher costs associated with pumping (Fig. 5.1).

A year earlier, the environmental regulator (the Environment Agency (EA)) had raised concerns over this practice and advised the company to undertake conjunctive use of reservoir and surface water sources so as to delay drawdown of reservoirs. The agency admitted, however, that operating the resource

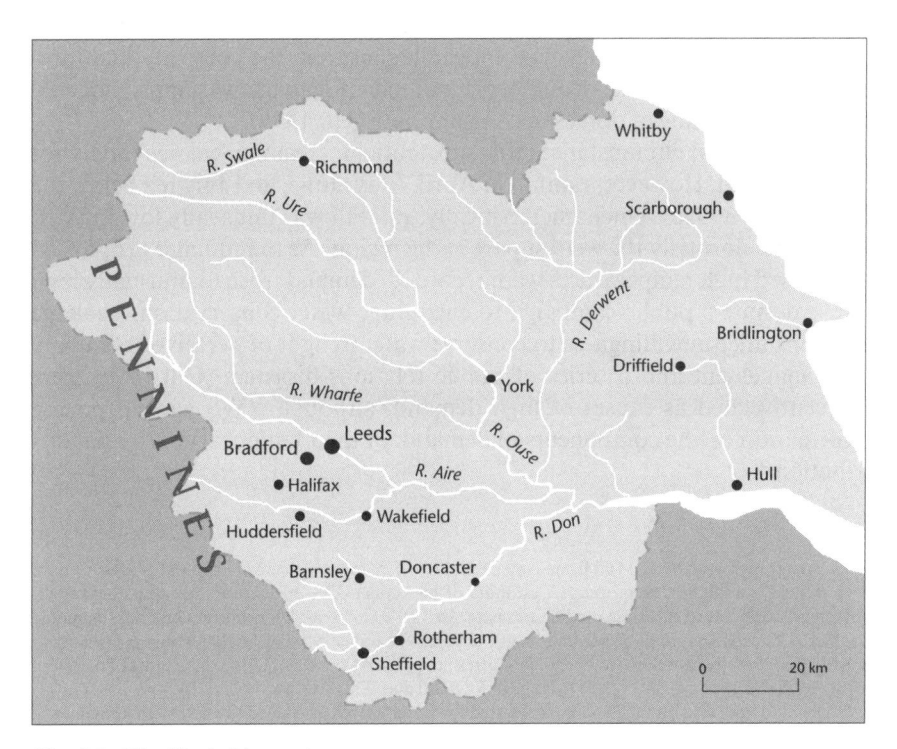

Fig. 5.1. The Yorkshire region.

system 'to obtain the maximum yield may be more expensive than operating to a lower yield because this may entail higher treatment or pumping costs' (NRA 1994b: 21). YWS began conjunctive use in 1994, to a limited degree, but there appeared to be no reason to deviate from standard practice. Demand forecasts predicted that water use would not rise greatly in the years following privatization; industrial demand was predicted to decline and domestic consumption to remain fairly flat. Although there was a fairly narrow margin between supply and demand, it was assumed that there would be no cause for concern until the end of the century (NRA 1994c).

The assumption that future demand would remain stable was perhaps one of the reasons why two key measures which would have increased supply system security—leakage control and headroom (the margin between supply and demand) management—received relatively little investment during early 1995. As Ofwat later noted: 'leakage problems were not being resolved at a sufficient rate, monitoring of leakage was not as comprehensive as it should have been, and priority placed on leakage control prior to 1995 was inadequate. The high rate of leakage progressively undermined the company's ability to keep within its distribution input forecasts . . . on any reasonable interpretation . . . leakage in [West Yorkshire] was significantly higher than that perceived by the company' (Ofwat 1996a: 2.7). With few meters installed at the household or even local level, YWS had scant information on distribution and amount of leakage (Ofwat 1996a: 2.6). Leakage was indeed higher than the company had forecast; at the height of the drought, the volume of leakage was approximately equal to the volume of domestic consumption[2] (Uff 1996).

Under normal circumstances this substantial increase in leakage might not have mattered. However, rainfall in April, May, June, and July fell below the LTA (NRA 1995c). Uncharacteristically, rainfall was unusually low in West Yorkshire, normally the wettest part of the region. As the summer progressed and record high temperatures were recorded, demand[3] rose to unusually high levels, despite a public campaign to encourage water conservation. Leakage and consumer unwillingness to conserve water in light of perceived company mismanagement and a series of public relations disasters (Gill 1995) were frequently cited as causes of high demand, although YWS had no precise information on the components of demand, or their temporal and spatial distribution.[4]

---

[2] Leakage increased by 46 Ml/d from 1989–90 to 1994–5. Various percentage figures for leakage have been quoted, the highest being the EA estimate of just under 37% of distribution input. Leakage is notoriously difficult to calculate, and figures expressed in leakage mask important variables, including number of households supplied per unit length of pipe. The industry accordingly favours measurement in Ml/d (million litres per day), or l/prop/day (litres per property per day). Using these units of measurement, YWS did not have the highest leakage in England and Wales (Ofwat 1997a).

[3] 'Demand' includes leakage and unaccounted for water as well as water delivered to customers, and is perhaps better captured by the term favoured by the regulators, 'distribution output'.

[4] Interview: Environment Agency, Leeds (December 1997).

Demand remained high throughout the summer, and reservoir levels fell rapidly. The most severe problems were experienced with the small reservoir sources supplying the cities of Bradford and Halifax and other communities in West Yorkshire. The small size of the reservoirs—typically, total storage in these reservoirs represented 120 days normal supply—rendered them inherently vulnerable to drought conditions (Uff 1996). In the dry summer of 1995 the reservoirs were not replenished by summer rainfall.

By July, levels in some reservoir groups in West Yorkshire had fallen to dangerously low levels. By August, several reservoir groups were at 25% capacity or less. As the Grid was set up to pump water in only one direction, from the normally wet west to the dry east, there was no mechanism available to supply water to areas needing it most (Uff 1996; NRA 1996a. YWS then imposed hosepipe (garden hose) bans in the west and north of the region, in which more than half of its 4.5 million customers live. The company also applied for drought orders[5] allowing increased river abstraction above licensed conditions, six weeks later than should have been the case under YWS' own guidelines (Ofwat 1996a). As yet, YWS had not declared an extensive leakage reduction programme, nor were major improvements to Grid flexibility planned. Instead, in the first of many public relations débâcles, the company continued to encourage customer conservation through a series of advertisements and a letter to business customers asking them to consider relocating outside of the Yorkshire region (*Yorkshire Post* 1995). The following month, the company sent a cheque for £10 to all domestic customers, a refund based on 'efficiency savings since privatization'. The 'dividend' to 1.8 million households cost the company about £20m., and sparked further public debate on company investment policy and water supply system management (Farrar 1996).

September rainfall was close to the LTA, and YWS held off on implementing emergency measures. But rainfall fell below the LTA in October and November, and reservoir recharge, which conventionally begins in the autumn, was not taking place. YWS realized it was facing ongoing difficulty with water supplies even if winter rainfall returned to average levels (NRA 1995b). The company announced that cuts in supplies would be necessary, applied for the necessary drought orders (Rofe and Hampson 1995a; 1995b; Nixon 1995), and erected trial standpipes in West Yorkshire; amid public outcry and vandalism of the standpipes, rota cuts were decided upon instead (Lazenby 1995). In an attempt to avert rota cuts, YWS had begun 'tankering' (hauling by large tanker trucks) water by road in September, filling tankers from rivers in east Yorkshire and even beyond the region in an attempt to refill

---

[5] As specified in the Water Resources Act (1991), water companies are required to apply for Drought Orders or Permits when a 'serious deficiency of water supplies' exists or is threatened by an exceptional shortage of rain. Drought orders cover fixed periods, and give licensed water undertakers the power to alter their normal conditions of operation—for example, to increase abstractions, reduce compensating flows from reservoirs, or to restrict non-essential use by domestic consumers.

reservoirs in West Yorkshire. At the height of the tankering operation, over 700 tankers were in use across Yorkshire, 24 hours per day. Tankering ceased in January as reservoir levels improved, and YWS withdrew its application for rota cuts (Barkas and Walker 1996).

The tankering operation cost YWS approximately £47m. (Uff 1996). Costs from infrastructure improvements also began accruing in the autumn; a £50m. package of emergency measures was announced in September, with an additional £50m. expenditure announced in October (Uff 1996). Five time-limited abstraction licences were granted, allowing the company to abstract water from the Ouse, Ure, and Wharfe rivers. The abstractions reduced flows in some areas to all-time lows with 'clearly detectable' changes to riparian ecology (EA 1996c). A further series of infrastructure works were implemented under Emergency Drought Schemes aimed at reducing leakage, increasing pumping capacity, and improving transmission links.

However, by February of 1996, reservoir recovery had still not occurred in west and south Yorkshire; with below average rainfall across the region, the Pennine reservoir groups remained at critical levels (NRA 1995c). Customers experienced problems with low pressure and poor water quality. In the same month, and to considerable public resentment, YWS announced an annual price increase of 5.6% in water bills for the year, the maximum allowable increase under the investment plan agreed upon by Ofwat (Buckley 1996).[6] Nonetheless, the utility's financial health did not appear to have been affected by the ongoing supply problems; in April, the company announced a better-than-expected 14% increase in pre-tax profits (Martinson 1996a), with profits of £162.2m. (Lazenby 1996) and forecast dividends rising accordingly. In the same month, drought orders were extended across the entire Yorkshire region, and the company announced a £70m. 'emergency package' to combat water shortages and 'win back its customers' confidence' (Boulton 1996: 4). Further improvements to the Grid were made, and despite concerns about environmental impacts, an emergency pipeline was built that could, if necessary, allow imports of water to Yorkshire from the wetter region to the north.[7]

In May, YWS applied for an extension of existing drought orders, and the Environment Agency (1996a) noted with concern that resources were even more fragile than at the same time the preceding year. The capital expenditure programme was expanded. By early summer of 1996, YWS had spent well over £100m. to lay 100 kilometres of mains pipes and to build 11 pumping stations (Bunyan 1995; NRA 1996b). An additional financial penalty arose in June as

---

[6] Allowable increases on the tariff basket of water charges are calculated using a price cap system of regulation for all privatized utilities (Vickers and Yarrow 1988). RPI $-$ X, X $>$ 0 is the standard formula, where RPI is the retail price index, and X represents other price factors, such as efficiency gains. Given the anticipated large capital expenditure required since 1989, the price formula for the water industry is calculated slightly differently, using the formula RPI + K, K $>$ 0. The K value negotiated by YWS in 1994 was 2.5%, and was intended to remain in place from 1994 to the next Periodic Review in 1999.

[7] The 13 km pipeline will form part of a larger water-transfer system crossing the watershed divide between Northumbrian and Yorkshire regions (MacGuire 1996).

Ofwat, in an unusual step, directed the company to reduce its price increases in subsequent years, costing the company approximately £40m. in revenue.[8] In addition, the company was ordered to provide compensation for homes or businesses whose water supply had been cut off,[9] and it also pledged to spend another £50m. on further reducing supply interruptions and leaks (Lascelles and Martinson 1996; Martinson 1996b; 1996c).

And then it began to rain—a relatively average summer rainfall for Yorkshire. With water consumption restrictions still in place, and with the continued licensing of increased abstraction possible from rivers, reservoirs slowly began to refill. By late autumn of 1996, supplies were close to 'normal', and YWS declared an end to the drought with the lifting of all restrictions on domestic water use.

Despite the fact that two independent inquiries into the drought concluded YWS had mismanaged water resources and supply, share prices remained healthy throughout the drought. This reflected the recognition that the company's revenues were not directly linked to performance, in the context of a monopolistic supplier whose prices, and thus revenues are set in advance and are, in the absence of widespread metering, independent of consumption levels. The sustained share price level was also in part attributable to the company's perceived ability to maintain dividends. The movement of capital from the regulated utility businesses to their newly established, unregulated parent companies had been the subject of scrutiny since privatization (Shaoul 1997). In the summer of 1995, YWS paid a £50m. special dividend to its parent company, Yorkshire Water plc. In its inquiry into the drought in 1996, the economic regulator of the water industry (the Office of Water Services) noted that 'in retrospect, the distribution of this special dividend to its parent company was not appropriate in the circumstances. The company [YWS] may need to borrow more than was assumed . . . or it may wish to consider asking its parent for a cash injection in order to finance its functions' (Ofwat 1996a: 1.10). Nevertheless, with low gearing and an underutilized balance sheet, YWS' parent company was still perceived to be a good investment. This was borne out in November 1996, as Yorkshire Water announced a 10% increase in interim pre-tax profits to £109.4m. (Martinson 1996d).

Producing scarcity

That YWS remained in good financial health during and after the drought is entirely consistent with the post-privatization regulatory framework. As explored below, there are two key factors in this apparent paradox: the cyclical

---

[8] Rate increases were to be limited to RPI in 1997–8, and then were to fall lower than the originally agreed-upon K value up to the year 2000 (OFWAT 1996a).

[9] £10/day for domestic customers, £50/day for businesses (Ofwat 1995b; 1996j).

investment patterns encouraged by the periodicity of economic regulation, and an incentive to underinvest characteristic of company strategy in the context of the regulatory 'game' played with the economic regulator in the first years following privatization. Taken together, these factors—given the 'lumpy' nature of water resources investment and large capital expenditure requirements—discouraged Yorkshire Water Services from maintaining existing infrastructure as needed.

Yet the drought was not merely a matter of investment strategies. After an exceptionally wet winter, the rainfall gradient in Yorkshire had 'flip-flopped', leaving the normally wet western part of the region abnormally dry. Accusations of mismanagement do not explain why the regulators of the water industry, as well as the water company, failed to anticipate low supply system security and associated vulnerability to drought, despite warning signals. Several trends, including the frequency of drought orders and hosepipe bans in the years following privatization, might have indicated that past climate trends were no longer reliable indicators of present patterns (Hughes 1996). Yet it was not recognized until 1994 that the regional water grid was insufficiently flexible to supply West Yorkshire (Uff 1996; NRA 1995c). Both the company and regulators failed to recognize that an overall supply deficit had existed within the Yorkshire region for several years prior to 1995. In 1995, this was exacerbated by supply system management practices based on demand forecasts that had consistently underestimated actual demand and by an increase in leakage to more than 30% of water put into supply, which was the 'key factor responsible for higher than expected [demand] . . . in 1995' (Ofwat 1996a). These factors, taken together, produced the crisis in the water supply system. If demand had remained stable, if the companies' own leakage reduction targets had been met, and had the improvements to the Grid been implemented even six months earlier, there would have been sufficient water to avert rota cuts and tankering (Uff 1996: 89). Yet, as the brief chronology of the drought demonstrated, the water company's management of the water supply system in the months preceding the drought undermined its ability to cope with the extreme demands placed upon it during the summer.

So why, and how, did YWS manage its way into the 'drought'? The management of the drought, as I detail in the following sections, was an outcome of a particular discursive construction of the performance of nature (weather) and consumers (demand), associated investment decisions negotiated through specific regulatory practices, and financial imperatives under which the water industry operated post-privatization. The following analysis details the reliance of water supply system managers on a discursive storyline positing climate as a stable phenomenon, the subsequent shaping of demand forecasts and investment strategies through the economic regulatory 'game', and changes in industry structure and corporate strategy following privatization and re-regulation.

Climatic stability: is the past a guide to the future?

The frequency and severity of drought-like conditions in the UK have increased in recent decades. The UK has experienced four major dry periods in the last twenty years: 1971 to 1976 (with 5 dry years out of 6), 1984 (February to August), 1988 to 1992, and 1995 to 1998 (Brown 1992*a*; EA 1998*e*; IOH 1995; Marsh, Monkhouse *et al.* 1994). Within this period, several riparian and hydrological systems set new records, fluctuating well beyond the average range of, for example, flow volumes (Arnell, Jenkins, and George 1994: 13). There has been a 'remarkable variability in hydrological conditions over the last decade' (IOH 1995; Marsh 1995). This recent volatility has led meteorologists to question whether 'the historical rarity of droughts [is] a reliable guide to their contemporary frequency' (Marsh 1995: 5) and whether current water management techniques are appropriate (Marsh and Turton 1996). This newly recognized volatility of environmental change 'tests the conventional wisdom that depicts nature as tending toward stability or near constant balance' (Zimmerer 1994: 110). This assumption, however, has until recently formed the basis for demand projections and industry design standards.

This, in turn, has implications for companies' approach to forcasting and drought management. Before 1995, for example, YWS's records ran from the 1930s to the mid-1970s (deliberately leaving out the 1974 drought), a relatively quiescent hydrological period. Since 1995, in a gradual process of updating and extending rainfall records, YWS has extrapolated rainfall data sets back to the 1920s, now including the 1929, 1933, and 1994 droughts. The company has also moved to a longer record period, using data from the 1970s to 1994, a 20-year record period which takes into account most of the major recent droughts.[10]

That YWS's hydrologists did not feel the need to update rainfall records in early 1990s is perhaps surprising considering that the company had applied for hosepipe bans and drought orders in nearly every year since privatization. Company standards specified that hosepipe bans should be applied no more than 1 in 8 years, and that drought orders should be implemented no more than 1 in 40 years, but hosepipe bans were applied for in 5 years out of the 7 following privatization and drought orders in every year but 1993. From the company's perspective, however, this did not entail a breach in design standards. Standards based on return periods imply nothing about the year-to-year recurrence of events, but express the average interval between events of a given magnitude, where that average is determined over a much longer period than that since privatization. Despite the finding of an independent inquiry that YWS did breach standards on both hosepipe bans and drought orders, 'and that these breaches are entirely consistent with the shortcomings both in the supply–demand balance and the distribution system' (Uff 1996: 7.26), some

---

[10] Interviews: Environment Agency, Leeds, and Yorkshire Water, Bradford and Leeds (December 1997).

YWS staff did not accept that they had breached design standards. From the perspective of company managers, unluckily, but unforeseeably, several 1-in-8 and 1-in-40 year events occurred in a relatively short period of time. Climate patterns were understood to be stable, and thus extrapolation to the present from past rainfall records assumed to be reliable.

This assumption of stability was critical to interpretations of rainfall data and drought severity by YWS, sanctioning acceptable and disallowing unacceptable data. The statistical method[11] used by YWS to calculate the severity of a drought, upon which the granting of drought orders depends, produced results indicating greater rarity for droughts in the early 1990s than the methodology subsequently adopted by the environmental regulator. Company managers defined the drought as 'exceptionally severe', and excluded evidence to the contrary; during the drought, for example, a rain-gauge reading that showed higher-than-average levels of rain was deliberately changed because it 'didn't look right' (Yorkshire Wildlife Trust 1996). With the assurance that droughts were 'severe', and with no evidence to the contrary, the yearly shortages in the water supply system could be ignored, and no supply deficit was perceived to exist in Yorkshire. This was perhaps part of the reason why YWS had added little reservoir capacity since the 1970s, relying mainly on increasing abstraction from rivers to increase distribution input and operating with the insufficient margin of supply over demand revealed in 1995.

Lies, damned lies, and demand forecasts

The assumption of a stable climate might not have mattered had demand not risen so sharply in 1995. This rise in demand had not been anticipated by YWS, in spite of the fact that distribution input had been 'consistently and significantly higher than the company had forecast over the period 1989 to 1995' (Ofwat 1996a: 2.5). Most critically, public perception of company performance played an important role in some consumers' refusals to respond to appeals to conserve water (Research by Design 1996). YWS discursive mediations of the drought failed to convince the public that the drought had not resulted 'from water agencies' technological mediation of the drought through their own unsustainable operations' (Nevarez 1996: 259). The 'tactics of hegemony' employed so successfully by water managers in southern California during its 1985–91 drought failed to convince Yorkshire consumers (Research by Design 1996; WaterWatch 1996), and the company's attempt to generate public consent by excluding the public from decision-making arenas backfired. Not only did YWS, as in the California case, attempt to insulate itself from interference by portraying itself and its strategies as 'technical' rather than 'political', YWS also acted decisively to exclude a customer advocate, backed by a number of large institutional shareholders, from its board of

[11] YWS used the Tabony Tables, a statistical method for calculating return periods of rainfall events (Tabony 1977a; 1977b).

directors. But whereas the discursive mediation of the recent Californian drought as an 'environmental crisis' served to justify expansion of the water supply system as 'drought insurance', YWS's discursive mediation of the drought failed to convince consumers, and motivated consumer *resistance* rather than acquiescence to appeals to conserve (Haughton 1998).

Calculations of the impact of public perception and consumer resistance were not figured into YWS's demand forecasts. In the years following privatization, YWS made use of central-value, average-weather demand forecasts, which used 'most likely' assumptions about population growth, life-style change, and economic activity in constructing demand scenarios, to which an empirically derived 'peak factor' was added to cover the possibility of extreme dry weather events. These forecasts were consistently, although only slightly, lower than actual demand.

In 1994, however, a new forecasting guideline was introduced and implemented across the industry (UKWIR/NRA 1995). Based on this methodology, YWS produced a series of upper/central/lower forecasts and dry/normal/wet forecasts, taking into account various economic as well as climatic scenarios. Based on assumptions about declining industrial demand, relatively flat domestic demand, and declining leakage rates, the 1994 Strategic Business Plan predicted a fairly dramatic reduction in demand, well below that of the early 1990s, and then a flat demand profile well into the next century (Uff 1996).

Understanding the shape of this demand curve requires reference to the functions performed by demand forecasts. Demand forecasts serve a variety of purposes: planning investment in the distribution system; forecasting revenue for internal purposes; and justifying resource development plans in the Strategic Business Plan submitted to Ofwat at the periodic Price Review. This last function is of particular importance. As discussed in Chapters 3 and 4, under the British model of regulation, the Environment Agency is responsible for water resources management and related functions (such as drought management). In the first years following privatization, however, 'arms-length' regulation implied that oversight should be favoured over intervention, leaving companies relatively free to decide rescource management strategies. In practice, however, expenditure programmes are negotiated between the utility and all three primary regulators. Although no clear hierarchy exists amongst the various regulators, the economic regulator wields significant power through its approval of company investment plans in the course of the price cap determination procedure. Although water resources are the responsibility of the Environment Agency, it is Ofwat that ultimately determines whether resource developments proposed by water supply companies will be financed and will thus proceed.

Understanding company resource investment behaviour thus requires a close examination of economic regulation, and in particular of the 'regulatory game' played between the regulators and water companies. Under the current price cap system of economic regulation, in which company business plans are reviewed and price increases awarded every 5 years, incentives for efficiency gains are provided, in theory, through the ability to make profits by achieving savings on forecast expenditure programmes. In practice, however, expenditure programmes are negotiated between the utility and the regulators in a 'regulatory game' (Melville 1994), in which companies and the regulators balance strategic and economic imperatives.

The first stage of the regulatory game involves negotiation between the regulators and the company over the appropriate level of capital investment over the next regulatory period, through negotiation over the allowed investments in each company's Asset Management Plan (AMP). Firms have several strategies that they may adopt, but in 1994 tended to inflate their projected costs since, by doing so, they maximized the predicted shortfall between their projected income and expenditure. This shortfall must be met through increases in water bills above inflation; thus, maximizing predicted revenue shortfall in turn maximizes the price increases the regulators will allow for a utility's basket of water tariffs. To the extent that higher cost predictions result in higher K values (and thus bigger profits and dividends), firms have an incentive to 'inflate the [value of the] asset base, to inflate CAPEX, to argue for a high OPEX . . . [and] to maximise the assumed cost of capital' (Helm 1994: 21).

This strategy may be understood as a variation of the Averch–Johnson effect, in which utilities subject to rate-of-return regulation inflate capital expenditure forecasts in order to maximize revenue, which is linked to an explicitly negotiated rate of return on capital (Averch and Johnson 1962; Helm and Yarrow 1988). The difference in the British case is that, although capital expenditure forecasts are inflated, British utilities are freer to underspend on actual capital programmes. Price cap regulation thus incentivizes operational efficiency gains in a way that American-style rate-of-return regulation does not, but fails to prevent the inflation of investment programmes (and thus inflation of prices) to which the American model is prone.

This finding has important implications for water resources management. Insofar as water companies have an incentive to select water management options that inflate CAPEX, and in the absence of regulations requiring demand management be implemented (as was the case in 1994), demand management is disincentivized as a water resource management option. Under the constraints of the regulatory game leading up to the 1994 Periodic Review, in seeking to maximize CAPEX, firms selected capital-intensive water delivery solutions and favoured, for example, proposals for new reservoirs (capital expenditure) rather than fixing leaks. As a senior World Bank official noted in an internal memo soon after privatization:

There is an 'unholy alliance' leading to heavy over-investment at the expense of the consumer, and this is an intrinsic feature of the UK system. What do I mean? There are two elements to this. The first is the setting of standards with no regard (or very little) for the costs and benefits . . . The second is that there is no clear incentive for anyone to scrimp on the costs incurred in meeting these. For all involved—the companies, consultants, contractors and equipment manufacturers—there is every incentive to be conservative and to over-invest, with the Asset Management Plan, in my opinion, the formal mechanism whereby this over-investment takes place... the bottom line of all of this seemed to me that environmentalists, 'Brussels' and the water industry were happily going to excessively high standards and over-investment to achieve those, with the bill being footed, of course, by the consumer. (Briscoe in World Bank/IFC 1991: 3)

Once the determination of the K value has been made, the second stage of the regulatory game begins. With turnover from customers' bills increasing at a predictable rate each year, providing a degree of security of income, companies have incentives to achieve savings on predicted expenditure programmes, through, for example, efficiency gains or delaying capital expenditure (Foster 1992; Helm 1994). Under a regulatory regime where prices are capped, most major efficiency savings in labour have been made, and little if any expansion of the consumer base is possible, the main sources for increased profit arise from cost cutting. Improving efficiency is one means of doing this, which WSCs have attempted with varying degrees of success, particularly through outsourcing[12] and reductions in the labour force.

Following the 1994 Periodic Review, reducing the provision for the renewals of the underground network was a critical strategy for increasing profits for many water companies.[13] This 'reduction' was a deliberate strategy on the part of the companies:

The reality is that the prices a company is able to charge are governed by the money it is authorised to invest, so that profits are driven by expenditure. The water industry gets a guaranteed return on capital, which has affected the investment programmes—there is strong evidence that the industry is seeking to spend more capital than it needs to, fails to invest, but still charges tariffs related to the large investments authorised. (Global Water Report 1997: 17–18)

In other words, licensed water suppliers are allowed to increase charges for water supplied on the basis of increases in CAPEX, not on the basis of OPEX. Thus, it is more attractive for companies to invest in a reservoir (which requires a large amount of capital, allowing a company to hike prices while its operating costs drop) rather than fix leaks (which increases OPEX, but which cannot be recouped directly through increasing prices). Ofwat, interested in keeping price increases down, encourages companies to devise low CAPEX, high OPEX solutions, in other words, to take up DSM; the companies, in the

---

[12] Outsourcing has occurred in two ways: spinning-off companies from the regulated businesses which, staffed by former employees, render the same services on a consultancy basis; or subcontracting non-core (for example cleaning, computer programming), and even core functions (for example pipe maintenance).

[13] The water industry has an unusually high ratio of profit to value-added, averaging 51% from 1985–95 (Schofield and Shaoul 1996). Although this ratio dropped slightly immediately after privatization, it has since risen again to pre-privatization levels as the provision for renewal of infrastructure has declined, suggesting 'that the possibility of improving profitability rests chiefly on the ability to reduce the provision for the renewals of the underground network' (Schofield and Shaoul 1996: 12).

annual and 5-year review of their investment plans and K values, argue for the opposite, new capital-intensive resource developments. Thus arises a contentious aspect of the regulatory game:

> There may be a structural [regulatory] disincentive for demand management. For some companies... demand management measures will be cheaper than new resource development. The companies could deploy demand management as a means of delaying investment just after the [Periodic Review]. But they are all water engineers, and like to submit for large CAPEX and this gives them a guaranteed return. Also, if they deploy demand management, and under-invest in CAPEX, they have to fear a possible Ofwat penalty, or at least lots of scrutiny. So the regulatory system gives [companies] an incentive to bid for a 'capital-intensive' solution. The incentive should be to choose the cheapest option, but because the regulator is watching the companies' CAPEX expenditure, which he doesn't want to be too high, [and also doesn't want to be] hoodwinked, demand management measures are less likely to be taken up.[14]

For the WSCs, allowable profits are based on a return on capital employed; maximizing profit entails maximizing capital expenditure and minimizing operating costs, which 'is why [WSCs] overstate their case to Ofwat, secure agreed tariffs to provide the return, and then fail to make the capital investment' (Global Water Report 1997: 16). Profits are 'maximised by minimising costs—both OPEX and CAPEX—provided the regulator does not intervene or claw back excess returns' (Helm 1994: 25). Several utility regulators have recognized that there exists an incentive under price-cap regulation to reduce quality.[15] Once K values have been set, a firm has an incentive to cut costs, thus lowering quality, so as to increase profits.

This underinvestment is cyclical.[16] Eager to increase dividends and report high profits in order to reassure investors and capital markets after the regulatory review, firms tend to 'bunch' investment towards the latter half of the regulatory period; uncertainty in the run-up to the Periodic Review results in, however, a sharp decrease in investment. The result to date has been a cyclical investment pattern, dropping after each Periodic Review, peaking in the middle of the Review period, and dropping again before the next Review (Miller-Bakewell, 1998; Ofwat 2002c).

Exploring the intricacies of economic regulation serves to highlight a number of salient points about the 'regulatory game', the most important of which is its cyclical nature. This cyclical effect is imposed by the structure of economic regulation, with its periodic reviews of water pricing. The intensification of regulation and information exchange in the run-up to the economic regulator's review of prices, overlooked by most assessments of the regulatory regime, has important effects on water resource management. This is of concern not only insofar as delays in investment may affect water quality, but also insofar as the infrastructure of much of the water network was already in a

[14] Interview: Environment Agency (July 1997).
[15] See, for example, the telephone regulator's first report (Oftel 1985: 13).
[16] I am indebted to Dieter Helm for this insight.

state of serious disrepair at the time of privatization (Foster 1992: 305). Furthermore, as water quality and water quantity are integrally interrelated, underinvestment in the latter can negatively impact on the former.

Prior to the Yorkshire drought, under the constraints of the regulatory game in the run-up to the Periodic Review in 1995, by maximizing forecast capital and operating expenditure and by minimizing forecast revenue, some companies maximized the predicted shortfall between their projected income and expenditure. This shortfall must be met through increases in water bills above inflation; thus, maximizing predicted revenue shortfall in turn maximizes the price increases the regulator will allow for a utility's basket of water tariffs. In 1994, demand forecasts were one pressure point in the regulatory game. Predicting a small increase or even a decrease in distribution output in 1994 allowed water companies to predict flat revenues. Given that the bulk of expenditure was expected to be on quality improvements, which entailed a large predicted capital expenditure, water companies asked for greater K values and thus greater price increases than would have been the case had demand forecasts matched the higher, actual demand curve in the 1990s.

With the awarding of K values, the second phase of the regulatory game began. With turnover from customers' bills increasing at a predictable rate each year, companies have incentives to achieve savings on predicted expenditure programmes, through, for example, efficiency gains or delaying capital expenditure. By 1995, YWS was slightly under its expenditure targets (Uff 1996).[17] The drawback to this two-phase strategy was revealed during the Yorkshire drought. YWS had not planned for what could have been an anticipated increase in demand; moreover, as a consequence of its own demand projections it had underinvested in water resources and infrastructure maintenance.

That this underinvestment, and the ensuing crisis in the water supply system, was not reflected in share performance is due to the structure of incentives within the regulatory framework, in which there is no formal link between revenue and performance. Revenues are linked to price caps, and set independently of performance, except for (light) financial penalties linked to a 'service standards' scheme set by Ofwat. Revenue and also share prices were thus effectively isolated from actual performance. Investors could thus remain confident that YWS—like the water sector in general—could continue to outperform the market, with the sector having shown a 56% increase (in real terms) in dividends since 1989 (Ofwat 1997b; OXERA 1996).

---

[17] Ofwat later publicly warned all water companies about the discrepancies between the investment programmes proposed in 1994—on the basis of which price increases were awarded—and their actual investment programmes (Ofwat 1996i).

Restructuring the water company: impacts on water management

The 'underinvestment' in water resources development by Yorkshire Water can only be viewed as such in hindsight. The capital programmes of, and resource development by, water companies are subject respectively to the approval of the economic and environmental regulators of the industry, the Office of Water Services (Ofwat) and the Environment Agency (EA). In the early 1990s, water resource issues were low on Ofwat's regulatory agenda; both regulators, in accordance with new European Union water quality directives, prioritized the amelioration of quality. Neither considerations of climate change nor planning with a margin of supply over and above predicted demand (planning with 'headroom') were allowed in Ofwat's 1994 Periodic Review; few submissions for resource expansion received approval.[18] Source control and leakage received little attention (Ofwat 1996h: 2.7). This focus was reflected in the relative proportion of capital expenditure on quality improvement, as opposed to resource expansion (only 5%), across the industry as a whole. YWS is typical of the industry in having spent 66% of its capital programme from 1990 to 1995 on water treatment, sewerage, and sewage treatment, and only 4% on water resources (Uff 1996).

In addition to quality improvements, corporate restructuring and diversification absorbed large proportions of managerial time and energy in the early 1990s. The spin-off of non-regulated firms in non-core, non-regulated areas of the business resulted in dramatic reductions in the labour force, continuing a downsizing trend initiated in the 1980s (O'Connell-Davidson 1993; Ogden and Glaister 1996). Like that of the majority of water companies, the management structure of YWS in 1995 differed significantly from its pre-privatization counterpart. Whereas the nationalized Yorkshire Water Authority had not been permitted to diversify beyond its core business, the privatized Yorkshire Water Services established a group of associated companies, only one of which was regulated, under the umbrella of a parent plc (Yorkshire Water 1995). The three non-core firms (an environmental services and engineering group, an overseas water and wastewater services contractor, and a property business) had either been purchased or hived off from the original water authority in a concerted effort to diversify the business both geographically and functionally.

Privatization also marked the beginning of significant restructuring within the utility. Until 1992, the basic unit of organization was geographical, based loosely on river basins within the region. Yorkshire Water was not unusual in this regard; many water companies were organized on geographical lines, reflecting the highly localized nature of the infrastructure systems inherited in

---

[18] Interviews: Yorkshire Water, Bradford (July 1996) and Ofwat, Birmingham (November 1997).

previous phases of rationalization of the industry and amalgamation of water providers. With the creation of the regional water authority in 1975, water resource functions had been consolidated with sewerage functions. By the late 1970s, there were seven operation divisions and a rivers division, which were subsequently further consolidated. The goal remained the allocation of responsibility to one manager intimately familiar with the water supply system in a defined geographic area.

This geographic division of labour was dismantled three years after privatization, with the establishment of a functional system of organization. Along with restructuring, both before and after privatization came downsizing, offering substantial cost savings for the company, given that labour was the single largest expenditure category for the water authority. From a peak of more than 6,000 employees in 1975, the staff was reduced to 4,900 in 1989, and further reduced to 3,300 in 1995 (Uff 1996); much of the later downsizing resulted from outsourcing of non-core functions such as cleaning and engineering consulting services (O'Connell-Davidson 1993), with a resulting loss of 'local' knowledge: 'Gone are the lovely days when I first started, when yes, I knew exactly what was happening on my little meter patch' (water-company employee, cited in Strang (2001: 60).

Water company restructuring rests on changing assumptions about nature of the water business post-privatization (Ernst 1994; Martin and Parker 1997). Only a subset of the nationalized water authorities' functions is still considered 'core' to the privatized water companies. Outsourcing, job shedding and restructuring have occurred across the water industry since privatization, along with a decisive change in management structure (Ferner and Colling 1993). Fewer managers are engineers who have spent their careers in the water industry. The employee profile of YWS managers had, for example, changed significantly since 1995, with more movement into the company of professional managers, 'people who know what the world is like outside a monopoly'. In 1994, management consultants were recruited, and mandated to examine YWS systems and identify possible cost savings and efficiency gains. Notably, major investments were made in information technology in operational, modelling, and assessment functions.[19] The subsequent corporate restructuring and diversification, while resulting in recognized efficiencies,[20] altered information flows and both actors and scales implicated in decision-making within the company. Functional specialization 'impaired the flow of information and hence [YWS'] ability to recognise the connection between issues in geographical areas' (Ofwat 1996: 7.3). Outsourcing further altered this flow of information. Leakage and pipe maintenance, for example, were outsourced to local contractors, with resulting cost efficiencies but with the loss of 'labour-intensive methods used up to the 1970s with locally based

---

[19] Interview: former Yorkshire Water manager (July 1996).
[20] The scale of these efficiencies is disputed. See, for example, Hunt and Lynk (1995).

teams which possessed a seemingly intimate knowledge of the system' (Uff 1996: 58). Poor pipe maintenance exacerbated the poor quality of the infrastructure in West Yorkshire, an area in which acute problems with corrosion had been recognized, without being redressed, since the mid-1970s (Pearce 1982: 58). With top managers preoccupied with information technology, restructuring, and diversification, 'there was little evidence of a management culture in which [resources] were at the top of the company's management agenda. Consequently the quality of decision-making [in 1995] appeared inadequate' (Ofwat 1996a: 7.2). By the time managers realized the seriousness of the supply-system crisis it was almost too late.

Re-regulating water management

Although the water supply system did not fail in Yorkshire in 1995, the drought brought issues of equity and access to the fore, largely through consumer resistance and the work of self-appointed and vocal water industry watchdogs. Criticism centred around the disjuncture between corporate and consumer interests, and between perceptions of water as a private commodity or public good (Letza and Smallman 1996; Haughton 1998). The terms and profile of the debate contrasted sharply with the relative invisibility of water resources prior to privatization, during which a 'flush and forget' attitude characterized the low profile, nationalized water industry (Maloney and Richardson 1994; Ward 1996).

Post-privatization, the expanded capital investment programmes of the private water companies were directed largely towards improving water quality after decades of perceived neglect. Environmental spend was thus the focus of attention in the first review period following privatization; water resources and associated infrastructure were relatively neglected. The underinvestment observed in Yorkshire was in fact characteristic of the privatized water industry as a whole. The production of the Yorkshire drought is thus compatible with, and to some extent the result of regulatory policies geared towards low-capital and high-revenue company investment programmes aimed at minimizing price increases to consumers. In this limited sense, the 1995–6 water shortage in Yorkshire was cost-effective, and the drought a least-cost solution to constraints on the hydro-social cycle.

In this reading, the Yorkshire drought was a form of 'governance failure' (Jessop 1995), emerging from contradictions in the uneasy after-Fordist institutional compromise between accumulation strategies of the newly privatized, re-regionalized water industry and the institutions binding the new, quasi-autonomous regulatory bodies. This new institutional fix contained within itself the seeds of the Yorkshire drought, rooted as it was in the contradiction between the duty to provide a public good and the desire to maximize profit, embodied in the dual identity of water as both merit good and commodity. Responses to the drought, such as the newly elected Labour government's 1997 Water Summit, the introduction of a windfall tax on privatized utilities, and companies' recalibration of source yields represent contested and, to a degree, experimental attempts to reach a (temporarily) stable, albeit internally contradictory, regulatory compromise.

In the aftermath of the drought, the water industry, regulators and government closely re-examined management of water resources (DOE 1996*d*; UK House of Commons Environment Committee 1996; NRA 1994*a*; 1995*a*; WSA 1997). Ofwat and the Environment Agency altered their water-resources strategy; for example, both climate change and 'headroom' were included in Ofwat's next Periodic Review (OXERA 1997). Public outcry over leakage rates has also forced the re-regulation of water supply management. With industry leakage rates averaging 150 l/head/d in 1996/7 (Ofwat 1997a), which the regulator estimated to lie between 25 and 30% of total water put into supply[21] (Ofwat 1996*a*), political and public pressure for leakage reduction was high. At its May 1997 Water Summit, the Government called for a regulator-led drought planning and management process.

The Environment Agency issued drought planning guidance to water companies in June 1999. The Water Summit's action plan for the industry also included mandatory leakage reduction targets, which Ofwat now monitors and updates annually (Ofwat 1997*f*). Ofwat now assesses company performance on the basis of 'water delivered' rather than 'water supplied', forcing companies to link operational efficiency to leakage levels.

Water companies have agreed to create water efficiency plans, and since early 1996, have a new statutory duty to promote the efficient use (i.e. conservation) of water by both domestic and non-domestic customers, and report to the economic regulator on 'water efficiency' indicators of customers' water conservation measures (OXERA 1997).[22] Companies also have a statutory duty to conserve water—in other words, to be operationally efficient—in carrying out their functions. The EA scrutinizes and encourages these operational efficiency measures, and to demonstrate that demand-management measures (leakage reduction, and metering installation) have been implemented, before granting new abstraction licences, as supported by enabling legislation[23] (DOE 1996*a*; EA 1996*a*; 1996b). At least one licence has been refused, to North-West Water, on the grounds that the company had not implemented sufficient conservation measures.

---

[21] Leakage is notoriously difficult to calculate, and figures expressed in leakage mask important variables, including number of households supplied per unit length of pipe. The industry accordingly favours measurement in Ml/d (million litres per day), or l/prop/day (litres per property per day). The percentage figure is supplied as a means of establishing relative magnitude of leakage only.

[22] The reporting requirements include indicators (for example numbers of [water saving] cistern devices), and companies are asked to provide a figure on the total volume and total costs of the water efficiency savings.

[23] Environment Act 1995.

Re-regulating water supply management: drought as an 'emblem' of crisis

The regulatory framework has thus undergone significant structural and normative transformations as a result of the drought. This process of re-regulation has been subject to intense political scrutiny and heated public debate, in part because re-regulation has departed from economic efficiency and market-based techniques, and reimposed command-and-control techniques. The Yorkshire drought figured within this debate as a symbolic event, repeatedly invoked by stakeholders in the debate over water supply management. In this way, the drought functions as an emblem, a metonym mobilizing bias; the perceived failings of YWS became the the 'issue in terms of which people understand the larger whole' (Hajer 1995: 20). Long after the end of the drought was officially declared, the drought-as-emblem has provided a pretext for the legitimization of innovation in normative codes and regulatory, management, and institutional frameworks. This observation about the political uses to which the threat of water scarcity is put has, of course, been made with respect to other regions and drought episodes (Allan and Karshenas 1996; Day 1987; Schramm and Kluge 1994; Solway 1994; Nevarez 1996; West and Smith 1996). In all of these cases, discourse about drought is much more than mere language—it is a matter 'of certain concrete discursive effects, rather than of signification as such' (Eagleton 1991: 223). In the Yorkshire case, its specific importance is its role in the ongoing transformation of the post-privatization regulatory framework. Ironically, although privatization was intended to remove companies from regular political interference and vigorous regulation, perceived company mismanagement and regulatory failings, together with a more open and participatory policy process, have resulted in greater public and government scrutiny (Maloney and Richardson 1995; Richardson, Maloney and Rudig 1992; Ward 1996). The Yorkshire drought has figured in this transformation as an emblem of a crisis in privatized water management, as 'a site where the established institutions of society are put to the test' (Hajer 1995: 39). The near-failure of the water supply system in 1995 represented, of course, a very material risk; yet the discursive mediation of the drought was central to its perception as management failure and to subsequent regulatory change.

Discourse analysis of the drought exposes more than merely political agendas shaped by competing interests. It is overly simplistic to argue that competing social interests give rise to competing policies or that dominant interests control expertise and hence shape the available knowledge to reinforce their agenda (Wynne 1994: 175–6). Discourse is more than a 'language of legitimation' (Hay 1995) or 'articulated perceptions' (Zimmerer 1993), or a means by which societal conflict is portrayed by the media and perceived by firms, governments, and interest groups (Day 1987; Haughton 1998; Nevarez 1996; Solway 1994). All representations of reality, not just narratives, are guided by discursive practices that mediate between our perceptions and reality.

Discourse analysis thus enables scrutiny of the processes by which nature is fixed in the human imagination; this in turn enables the establishment of an analytical link between the material and social dimensions of resource regulation. Discourse analysis of regulation thus enables an exploration of the creative, seemingly chaotic transition from one mode of social regulation to the next (Bakker 1999a; Hay 1995). Focusing on storylines rather than narratives of crisis highlights re-regulation as a transformative process, raising questions of political legitimacy and public perceptions as well as of the nature of socio-economic restructuring. Socio-economic restructuring, in other words, entails the re-regulation of a dynamic resource landscape (Roberts and Emel 1992); this re-regulation is, in part, a discursive process, with very real, material effects. Addressing the ecological dimension of socio-economic restructuring thus requires a scale-hopping exploration of the intricacies of 'real' regulation, and their interplay with macroeconomic trends. This implies a move away from analytical reliance on the concept of crisis, with its underlying assumption of a rupture or distinct break in economic practices and social life. This is not to deny that crises occur, but to warn of overemphasizing their role in accounts of historical-ecological transformation.

Conclusions

The restructuring of YWS post-privatization entailed changing flows of information and reconfiguring of resource- and quality management techniques, altering not only decision-making practices but also the sanctioned actors and information involved in decision-making. Discourses of climatic stability and demand forecasting produced a set of understandings about the environment and consumption patterns that simultaneously concealed or even rejected contradictory evidence—the warning signals of the early 1990s. Operating under the assumption of a stable climate and (supposedly) sufficient resources given present demand, water resources issues were considered by neither the regulators nor the water utilities to be as critical as the task of improving water quality and corporate efficiency. The drought was authorless, unintended, yet nonetheless had a kind of economic intelligibility; it was, in a limited sense, the least-cost option for YWS given the constraints of the post-privatization regulatory 'game'. The Yorkshire drought was neither simply a freak of nature, nor an isolated case of spectacular mismanagement of a water supply system, but rather what Neil Smith terms 'produced scarcity in nature' (Smith 1984: 60). Critical to this production of scarcity was the absence of any formalized link within the economic regulatory framework between company performance and company revenues, as witnessed by Yorkshire Water's record profits and healthy share performance during and after the drought. Equally critical was an apparent inability on the part of the environmental regulatory framework to anticipate and/or respond to structural weaknesses within the water supply management system.

The structure of the regulatory framework was thus of concern to government and regulators following the drought. A central question raised prior to privatization of the water industry was whether, given long lead times and substantial capital requirements, private companies would be able to provide a sufficient level of resource and supply security for consumers. Several key changes accompanied privatization: the shift in water management regulation and practice from a prioritization of security of supply to a prioritization of cost-recovery; a shift in emphasis from water quantity (resources) to water quality (environmental spend) in capital expenditure programmes; a regulatory framework which encouraged cyclical capital investment through the Periodic Review cycle, combined with the incentive to minimize expenditure on agreed-upon infrastructure improvement programmes; and a focus of key regulatory actors on *economic* security of companies rather than security of supply in engineering terms. The combined results of these changes were necessary conditions for the Yorkshire drought, and the substantial re-regulation that has taken place in the industry since 1995 can be interpreted as implying a belief, on the part of the government, in the existence of structural flaws in the post-privatization regulatory framework, and as resting on a recognition of political responsibility for ameliorating those flaws.

The substantial 'regulatory creep' that has resulted has undermined earlier visions of 'regulation with a light touch' that would 'wither away' with time. The government has been again drawn into active political role in regulating water—as borne out by the 'quadripartite' *ad hoc* regulatory process created at the 1995 Periodic Review, in which DEFRA plays a key role in adjudicating competing regulatory priorities, by the 1997 'Water Summit' at which the government set down performance criteria for the water industry, and by subsequent periodic political pressure by ministers on regulators. More than a decade after privatization, the management of water in England and Wales is best characterized not as a process of deregulation, but as a process of *re*-regulation. The long-term implications of this trend are considered in Chapter 7, prior to which we continue our examination of the implications of privatization for domestic consumers in Chapter 6.

6

Thirsting for Equity: Consumers and the Contested Politics of Water Pricing

Over the past three decades, the progressive commercialization of water management has gradually transformed consumers' entitlements to water supply services in England and Wales. The business of domestic water supply, like other utilities, has been gradually evolving from the supply of a service to citizens at subsidized rates, towards the sale of a commodity to consumers on a full cost-recovery basis. This chapter analyses the implications of this ongoing process for domestic water users. The first section contrasts concepts of economic and social equity, details the inter- and intraregional equalization that was characteristic of the early phase of nationalized management of the water industry, and traces the evolution of policy and regulatory priorities in water charges to domestic consumers over the same time period. Two specific developments are considered: the shift from policies promoting geographical equalization to those resulting in differentiation of water rates; and the trend away from rateable values (intended as a proxy, however poor, for wealth) and towards metering (i.e. charges related to consumption). The analysis does not measure relative efficiency of water companies,[1] but rather focuses on the implications for consumers of the substitution of the yardstick of economic efficiency for the goal of equalization.[2] The political implications of the associated displacement of a principle of social equity by a principle of economic equity in water charging are considered in the second section. In the third and final section of the chapter, the concepts of justice implicit in the recent re-regulation of the water industry are analysed.

---

[1] Under the Keynesian regime, national ownership and subsidies to the water supply system were understood as vital subsidies to the private sector that contributed to both growth and stabilization of demand (Clarke 1993: 208; Gibson-Graham 1996: 181). In the 1980s, however, 'redistribution in the direction of social equity was viewed as a drain on potential investment' (Gibson-Graham 1996: 181). The fact of these differing goals confounds comparative assessments of the efficiency of the public and private sectors (Lapsley and Wright 1990).

[2] See O'Connell-Davidson (1993) and Saunders and Harris (1994) for analysis of the implications of water privatization for employees and shareholders.

Economic equity versus equalization

On the one hand some people say that water is so important that it should be free. On the other hand others say that it is so important that it is worth paying for ... We have a duty to collect charges from all those receiving our services ... we cannot reduce or abate charges on grounds of age, ill-health, low income or on any other such socio-economic criteria. (Jeffrey Phillips, Welsh Water, cited in AMA 1993)

The current prioritization of economic equity stands in distinct contrast to the principles of social equity underpinning water charging policy early in the nationalized era. The principle that public ownership of and subsidies to the water supply system were necessary—because of the multiple market failures to which water supply (most importantly, monopoly) is subject, and also because of the positive externalities associated with water and sanitation services—had been widely adopted at the turn of the century (Hassan 1998). Charges for water were not determined solely by the necessity of cost-recovery, but set in light of the goals of universal provision and equitable access (Sleeman 1953), implying a concept of social equity in which consumers paid according to their means rather than according to the costs they imposed on the system. Water was regarded as a strategic resource, and water charging policy was administered accordingly (Swyngedouw 1989; Graham and Marvin 1995). Charges for water were not determined by the necessity of cost-recovery, but set in light of the goals of universal provision and equitable access (Sleeman 1953).

The commercialization of water pricing policy has entailed a gradual move towards 'economic equity'—the principle that users of a utility service should pay, as near as possible, the costs they individually impose on the system (the 'benefit principle'). In contrast, the principle of 'social equity' underpinning earlier water policy implied that users should be charged according to their ability to pay (the 'ability-to-pay principle'). Throughout much of the twentieth century, charges were made and water networks extended on the principle of universal provision, with resulting cross-subsidization between 'lucrative urban areas and trunk networks and those groups that were relatively expensive to serve: rural areas and socially disadvantaged communities' (Graham and Marvin 1994: 114). From 1974 onwards, sewerage (and also water, if this was not already the case) bills were charged separately to the consumer, rather than through local taxes, as had been the case for the majority of consumers. The rateable value of property was the base employed to calculate these water and sewerage charges, as domestic properties were not metered. The employment of rateable property values as the base for water charges implied *intra*regional rate equalization (implying cross-subsidization between different classes of consumers within the RWA area); the existence of cross-subsidies was tolerated as water supply was regarded as an essential public health service and it was considered equitable for the costs to be borne according to ability to pay. The assumption was made that the value of a family's property was a good proxy for its wealth, despite the fact (as was recognized at the time) that property values are very imperfectly related to family income (Whiteley 1977, Lingard 1974).

A concept of social equity similarly underpinned interregional equalization measures, which were initiated soon after the creation of the RWAs in 1974. Underlying *inter*regional equalization—active cross subsidization between regions and/or classes of consumers—was the principle that users in different regions should pay, as near as possible, equal bills regardless of the cost they impose on the system, thus 'extending to the national level the process of the equalization of charges which [had] been . . . occurring within the areas of individual water authorities' (Frankham and Webb 1977: 197). Following the creation of the RWAs, two regions—Southern and Wales—had completely equalized prices, and the others had implemented varying degrees of equalization. Concerned about the disparities in average water supply charges *between* RWA regions (from 17% below to 44% above the national average (Porter 1978)) the Labour government initiated a review of the water industry in England and Wales (DOE 1976) which recommended equalization of charges between regions via direct transfers between water authorities and companies.

Following the review, the Water Charges Equalization Act was passed in 1977.[3] Under its provisions, the Secretary of State was empowered to order those water undertakers whose financing costs (depreciation plus interest of assets in use for water supply purposes) were less than the weighted industry average to pay an 'equalization levy' to the National Water Council, and to order the National Water Council to pay 'equalization payments' to undertakers whose financing costs were greater than average. This implied rate rises in regions where domestic water bills were cheaper, and rate reductions in regions where bills were more expensive.[4] As the Director of Finance of the administering body, the National Water Council (created by the Water Act of 1973 which also created the RWAs) noted, 'the objective of the [equalization] scheme is to reduce (but not eliminate) the disparity between average household bills for water supply in the areas of the ten water authorities . . . and rural areas . . . receive a subsidy at the expense of dense urban areas' (Porter 1978: 3). The equalization payment/levy was applied to all 29 of the statutory water supply companies, as well as the RWAs, from 1978 to 1981. In 1978/9, the total equalization payment/levy for the industry was just under £9,583,000; in 1979/80, £9,405,000, and in 1980/1, £9,832,000.[5] The amounts were relatively small—of the order of 2% of the industry's income from unmeasured water, but had in some cases a significant impact on water bills (Tables 6.1, 6.2).

[3] The 'Water Charges Equalization Act', Public General Acts & Measures, 1977, c. 41, came into force on 20 Oct. 1977. The Act was 'to provide for the payment of equalization levies and payments by and to statutory water undertakers in England and Wales' (s.1(1)). A statutory instrument was required each year levies and payments were to be made.

[4] Section 4(2) (a) and (b) of the Water Charges Equalization Act specify that 'the amount of any equalization levy and . . . any equalization payment which a statutory water company is entitled to receive in respect of any year shall be passed on in full, in the form of increased or reduced charges, to the persons to whom water is supplied by the company on an unmeasured basis in the corresponding accounting period.'

[5] S.I. 1977/2165; S.I. 1978/1921; S.I. 1979/1754.

Table 6.1. Equalization payments/(levies) and equivalent income, selected companies (1978/9)

Water authority/ Company	Equalization payment/(levy) (£)	Equivalent income per m³ of water supplied	Average domestic bill 1977/78 (£)[1]	Average domestic bill 1978/79(£)[1]	% change
Welsh WA	3,486,000	0.89	25.40	22.76	-10.4
Yorkshire WA	(375,000)	(0.08)	17.66	20.64	16.9

[1] Not corrected for inflation
Sources: DOE (1979), S.I. 1977/2165 and CIPFA (1979)

Table 6.2. Water authority average domestic bills, 1977/8–1978/9

Water authority areas	Average domestic bill 1977/78 (£)	1978/79 (£)	% change
South West	23.12	24.11	4.3
Welsh	25.40	22.76	-10.4
Anglian	19.96	18.80	-5.8
Wessex	19.28	23.63	22.6
Southern	17.40	19.00	9.2
Yorkshire	17.66	20.64	16.9
Northumbrian	17.56	17.67	0.6
North West	16.90	18.94	12.1
Severn Trent	15.71	16.17	2.9
Thames	15.02	17.85	18.8

Not corrected for inflation
Source: DOE (1979)

The Equalization Act did not provide for the payment of direct subsidies (via central Government grants) to RWAs with above-average charges, as the government was committed to phasing out subsidies for nationalized industries. The Labour government was nonetheless determined to address regional disparities and 'unreasonably' high costs in some RWA areas in accordance with 'equity and fairness'—implying a concept of *horizontal* social equity, in which all consumers should pay, as near as reasonably possible, the same charges for water services, despite the different costs they impose on the system (Frankham and Webb 1977: 198). Direct transfers between water companies and the resulting 'rebalancing' of water charges were thus employed as an instrument for the achievement of income-distribution objectives, despite evidence that rateable value-based charges were a poor proxy for income.[6]

In opposition, the Conservatives objected to the equalization scheme, arguing in favour of economic equity:

---

[6] The Layfield report, for example, indicated that rateable values were an imperfect measure of household income (Report of the Committee of Enquiry into Local Government Finance. HMSO Cmnd. 6453, 1976).

There may be, arguably, a case for once and for all equalization of the historic costs that each of the water authorities are carrying in various degrees. But the best principle is that those who use each water authority's services should pay the true cost of providing them. Raising costs in an area in an attempt to keep them down in others is the road to financial irresponsibility. Cross-subsidising one WA from another would also involve Treasury oversight and political interference by Department of the Environment Ministers. Worst of all, it would destroy the direct accountability of the RWAs to their own consumers for the true costs they incur. (Griffiths 1976: 4)

Following the election of the Conservative government in 1979, the equalization scheme was suspended. Critics of the scheme had argued that the pursuit of social equity, via equalization, was occurring to the detriment of efficiency, as equalization payments decoupled revenue and prices from costs, encouraging over-provision and over-consumption. Whilst attempting to equalize costs for consumers, in other words, the equalization policy was seen to discriminate against the RWAs that controlled their costs, and to benefit those that did not control their costs, removing incentives to contain costs and leading to higher than necessary prices overall (OECD 1987). Equalization with respect to capital charges, it was argued, penalized those companies with low capital charges in favour of those with high capital charges (due to a combination of lower population densities, more difficult terrain and, in some cases, heavier debt burdens).

The cessation of equalization payments was one of many reforms to the water regulatory framework throughout the 1980s. Beginning in 1981, the government employed powers available to it under the Water Act of 1973 to require the RWAs to achieve individualized rates of return on assets on a yearly basis, independent of their financing needs.[7] The justification, according to the minister then responsible for the industry, was that 'investment in the public sector must earn a return comparable to investment in the private sector' (King 1980). Instances of high increases in the required rate of return (particularly from 1985 to 1987) effectively decoupled charges related to capital assets from capital expenditure. This 'had the effect, according to authority declarations largely accepted by Government, of increasing charges at double the rate otherwise necessary' (CIPFA 1987: 7); by the late 1980s, water bills were increasing above the rate of inflation in several RWA areas. The government controls continued through to 1989, constituting not only a tool of fiscal discipline of the water companies, but also, as some water authority chairmen argued, an instrument of soft taxation in the case of profitable RWAs.

The regulatory framework created at privatization consolidated and formalized the move away from intraregional equalization of charges. Ofwat is charged with the duty of ensuring that 'there is no undue discrimination' (Water Industry Act 1999, s.2.3.a.ii) in the setting of charges for water and sewerage services; the regulator 'interprets this to mean that, where possible,

---

[7] Under S.I. 1981/826.

there should be no cross subsidy between classes of consumer' (Ofwat 2000a: 23). In other words, cross subsidies between services provided to recognized classes of consumers (for example from water to sewerage customers, or from industrial to domestic), as well as between individual consumers, should be eliminated. In contrast to the early phase of the nationalized era, when discrimination was understood to occur when customers were charged a different price for a technically similar product (i.e. a unit volume of water), discrimination is now implicitly understood to occur when customers are charged the same price for a product that has different supply costs in different regions. In practice, balanced by the duty to ensure 'that the interests of customers ... in rural areas are protected', this duty has been interpreted to mean that each customer should pay, as near as is practicable, the *actual* costs they impose on the water and sewerage systems. In adopting a definition of economic equity, Ofwat supports widespread (although not full) penetration of meters into domestic properties, whilst arguing that any regressive burden of water charges should be met through the social security system, not through corrective measures applied via water charging mechanisms.[8]

Implications of price rises for consumers

The application of the principle of economic equity undermines a practice common in most industrialized countries throughout much of the twentieth century, in which water regulation explicitly incorporated various social policy goals such as income redistribution, employment generation, and regional equalization (OECD 1999). The consequences of a shift from a policy goal of equalization (underpinned by a notion of social equity) towards one of efficiency maximization, and its corollary, neoclassically defined economic equity, are illustrated in Figure 6.1. Consumers in different regions have experienced significantly different rates of increase in charges for both water supply and sewerage services, in both relative and absolute terms (see the Appendix for a detailed analysis). Those regions with large lengths of polluted coastline, for example, such as the South-West region, have experienced increases in sewerage charges well above average, to meet increasingly stringent EU beaches and bathing water quality legislation. On both a relative and absolute basis, bills have increased much more quickly for consumers in these areas. The average unmeasured domestic water and sewerage bill in the South-West region in 1999/2000, for example, stood at £390 per year, as compared with a weighted industry average of £277 and the lowest average regional charge of £208 for those served by Thames Water. In two regions charges have also been 'de-averaged' regionally, by dividing the region into several zones where charges

---

[8] Ofwat commissioned an extensive study on the distributional aspects of various methods of charging households for water in the early 1990s (see Rajah and Smith 1993).

may differ by up to 10% of customers' bills.⁹ This has resulted in higher absolute intraregional differences in zone charges. Other water companies have not moved to regionally de-averaged charging in part because of the complexity involved in billing, but also because of the implications for rural consumers (i.e. higher prices); de-averaged charging schemes would likely conflict with Ofwat's duty to protect rural customers (Ofwat 1999*b*).

A key driver of the increase in water charges above RPI across the industry has been increased capital expenditure in the industry. Higher drinking water quality and lower environmental impact of water production have been two important resulting improvements. River water quality in Britain appears to be at its highest level since the Industrial Revolution, according to the government's environmental regulator (EA 2001; DEFRA 2001*a*). Drinking water quality has also improved during the decade following privatization, with only 1 in 500 tests failing to meet requirements—as opposed to 1 in 100 in 1990. Compliance for sewage-treatment works is up to 98.6% in 2000 from 90% in 1990. These improvements have been facilitated by more stringent water quality legislation, and tougher enforcement, expressed in the significantly higher number of prosecutions, which remain some of the most heavily prosecuted polluters in Brtitain (Fletcher 2002; EA 2003).

These improvements have required large capital expenditure programmes. The distributional implications of the associated increase in water bills have been widely debated (see for example NCC 1993; Thackray 1995; WHICH 1996). Some consumers—for example those with low consumption in a high-rateable value property—will see their bills drop when a meter is introduced. Others—large families in low-rateable value properties—may see dramatic rises in water and sewerage bills. In one study, low-income families were found to spend an average of about 4% of their weekly budget to pay for water (SCF 1996), significantly higher than the national average of just over 1%. Bills have risen above the rate of inflation throughout the 1990s (Fig. 6.2); the proportion of income of lower-income families spent on water and sewerage has risen faster than that of higher-income families (Drakeford 1997). While bills were rising quickly in the early 1990s, Ofwat suggested that the proportion of a single pensioner's income required to cover the cost of combined water and sewerage bills could be as high as 14% in some regions by 2004/5 (Ofwat 1993*b*).

Rapid price increases have occurred in other European countries, such as France, where large expenditure was required to meet increasingly stringent European water quality legislation (OECD 1999); insofar as price rises in England were necessitated by quality improvements, they should be regarded as independent of privatization. However, in England and Wales, rapidly increasing water charges were in some cases exacerbated by the price differential between non-metered and metered domestic customers, the proportion of

---

⁹ Severn–Trent and Thames Water charge on a zonal basis, and other companies have different charging areas as a result of company mergers (for example South-East Water, Sutton and East Surrey water supply companies) (Ofwat 1990: 121).

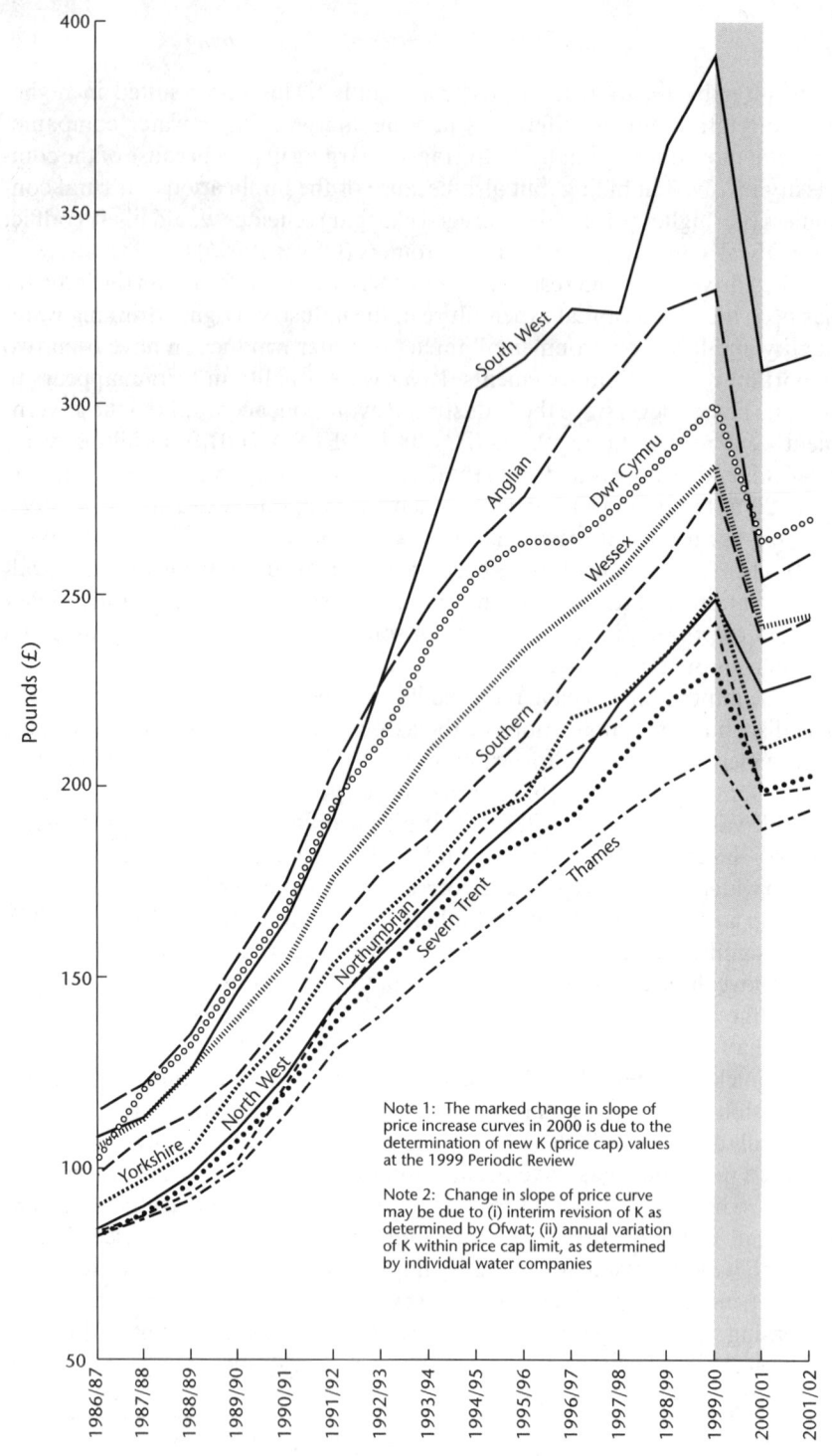

Fig. 6.1. Water and sewerage charges, by WaSC region (1989–2000).

Note 1: The marked change in slope of price increase curves in 2000 is due to the determination of new K (price cap) values at the 1999 Periodic Review

Note 2: Change in slope of price curve may be due to (i) interim revision of K as determined by Ofwat; (ii) annual variation of K within price cap limit, as determined by individual water companies

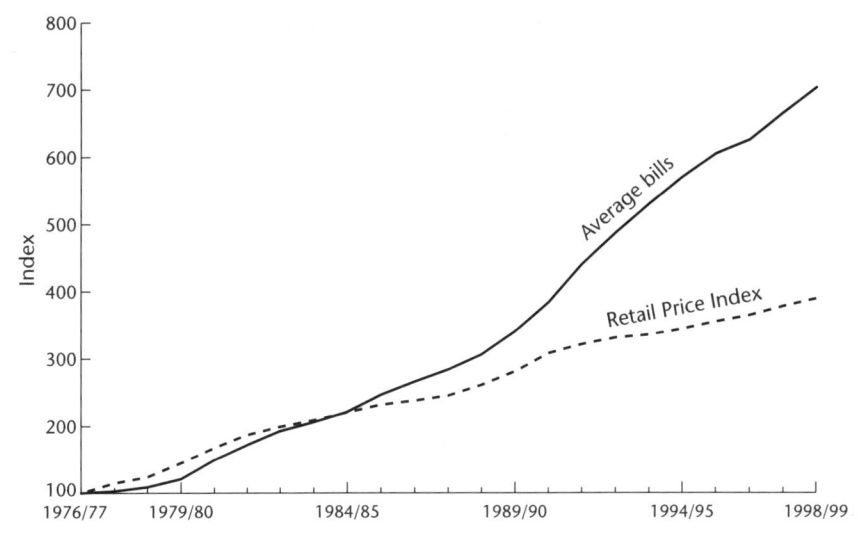

Fig. 6.2. Average domestic water and sewerage services charges (1976–1999).
(Data source: WSA (various years))

metered properties having increased significantly with the metering programmes initiated post-privatization. By the mid-1990s, some domestic consumers with metered supplies were paying—in some cases significantly—more per unit volume of water than consumers with non-metered supplies (SCF 1996; WHICH 1996). The proportion of the population with metered supplies has grown significantly over the 1990s (Table 6.3).

Water poverty

In response to NGO campaigns such as that of 'Save the Children', following a well-publicized study of the often health-endangering measures taken by low-income families to conserve water (SCF 1996), Ofwat has required companies to equalize the rates charged to metered and non-metered customers. The regulator now requires companies to maintain a differential of no more than £30 between measured and unmeasured domestic consumers (Ofwat 1999*a*). This step was also encouraged by research carried out by the British Medical Association, examining the links between household water disconnections and the sharp rise in reported dysentery rates in the early 1990s (Table 6.4) (BMA 1994).

The correlation between disconnections of domestic properties for non-payment of bills and the sharp rise in dysentery rates is not a simple one, as there are many confounding variables. Nor was the rise in disconnection rates simply attributable to water companies' more draconian policies. The

Table 6.3. Measured versus unmeasured water supply, households, 1989–1998

	Unmeasured (000)	Measured (000)	% measured
1989/90	18,771	147	0.78
1990/91	19,012	358	1.88
1991/92	18,934	523	2.76
1992/93	18,742	637	3.40
1993/94	18,871	890	4.72
1994/95	18,771	1,087	5.79
1995/96	18,711	1,330	7.11
1996/97	18,517	1,640	8.86
1997/98	18,087	2,268	12.54
1998/99	17,607	2,992	16.9

Source: WSA (1989 – 1998, 2000)

Table 6.4. Disconnections of domestic properties for non-payment of charges, 1984–1998

	89/90	90/91	91/92	92/93	93/94	94/95	94/95	95/96	96/97	97/98
Water service co.	5,747	5,417	13,234	11,141	6,916	5,115	5,115	3,169	1,833	1,205
Water supply co.	2,679	2,456	8,048	7,495	5,536	4,932	4,932	2,657	1,315	702
Total	8,426	7,873	21,282	18,636	12,452	10,047	10,047	5,826	3,148	1,907

Note: Disconnection of domestic consumers is now prohibited under the Water Industry Act (1999)
Sources: WSA (various years), Ofwat (personal communication)

cessation of payment of water bills via local authority rates, and lack of allowance for the rapid increases in water bills during the 1990s in Income Support due to changes in the benefits system, were factors in increased consumer water debt in the 1990s (Herbert and Kempson 1995). Water companies had been directed by Ofwat shortly after privatization to find alternative payment strategies for consumers (1991*b*), but disconnection rates rose nevertheless, in part due to the fairly widespread conviction within the industry that some level of disconnections was necessary as a means of demonstrating companies' willingness to force customers to pay water bills. During 1994 alone, almost two million households in Britain defaulted on water bills, and by the end of the year more than one million (5%) were behind with their payments (Herbert and Kempson, 1995). Another survey found that 75% of those on Income Support have difficulty paying water bills, and that water debt was rising faster than any other component of debt for low-income families (Marvin, Graham *et al.* 1996). Herbert and Kempson (1995) found that low income (as opposed to increases in water bills) was a significant factor in explaining water debt, with more than half of all households in water debt living in either local authority or housing association accommodation, and water debt being more common in the North and Midlands than elsewhere.

During the 1990s, disconnection rates and 'water poverty' became the source of much negative publicity for the water industry (Graham 1997; Harrison 1996; Huby 1995; Huby and Anthony 1997). The alleged shortcomings of the water industry were a focus of public health campaigners (see, for example, Middleton and Saunders 1997), consumers groups (for example the National Consumer Council, Consumer's Association), as well as the Labour opposition, in particular through the campaigns of then-shadow Labour health secretary Frank Dobson (Dobson 1995*a*; 1995*b*). Water companies were directed by Ofwat to reduce disconnection rates, which have fallen since peaking at over 20,000 properties disconnected in 1992 (representing one disconnection for every 1,000 households billed (Graham and Marvin 1994).

The reduction in disconnection rates is also partially attributable to the introduction of 'Budget Payment Units' (BPUs) (Marvin and Guy 1997*b*; Marvin, Chappells *et al.* 1998). These payment systems, consisting of a meter, together with an electronic 'smart' card, are installed inside individual customer's homes; 33,000 households had water BPUs installed by 1998 (Bannister 1998). The consumer recharges the card (most charging outlets are in post offices), and credits are transferred to the control box once the card is inserted. Each payment purchases a volume of water, or an amount of time during which the consumer can use the water service. When the credit runs out, an emergency period allows for additional use, at the end of which the water supply is stopped. As 'Ofwat does not believe that the customer's operation of the unit in this way amounts to action by the water company to cut off the customer's supply' (Ofwat 1996*f*: 2), these 'self-disconnections' are not calculated in the statistics of disconnection rates. National water disconnections fell from 10,047 to 5,826 in 1995/6, whilst the number of installed BPUs grew from a few hundred to 15, 077 (Harrison 1996). Two-thirds of households using water pre-payment meters incurred 'self-disconnection' in the first year of use[10] (*Utility Week,* 9 Sept. 1996; 6). Many of the units were installed as an option offered to low-income households with a history of non-payment of bills as a means of managing their outstanding debts to their water company (Graham 1997). Debt repayments are, in these cases, facilitated by the automatic deduction from the smart card towards the cost of debt repayment, as well as standard charges for the water service. Simultaneously, relatively high administrative costs posed by these low-profit marginal consumers (due to a combination of low consumption and high debt) are minimized by the water companies.

---

[10] BPUs were declared unlawful by the High Court in March 1998. Accordingly, no new BPUs will be installed, but 'companies will allow existing customers with units to continue to use them as a way of paying their bills' (Ofwat, News release, 12 Mar. 1998). The number of BPUs in commission has been steadily reducing since 1998 (Ofwat, personal communication).

The politics of equity

The rapid rise of water and sewerage charges since 1989 has occurred largely to pay for 'dirty' rather than 'clean' water services; as a result, water quality (of both treated water and surface and sea waters to which effluent is discharged) has improved over the past decade. The majority of expenditure has been directed towards improving drinking water quality and minimizing environmental impacts, rather than towards resource development. In contrast to the energy sector, where the UK's global warming obligations are not enforced by legislation, environmental expenditure in the water industry is driven by EU environmental legislation, in which water is among the issues most comprehensively addressed.

In privatizing the industry, government ministers believed that they had divested themselves of political responsibility for the rapid price increases associated with the high projected capital expenditures (estimated at £28bn. in 1989 prices), necessitated by the increases in standards imposed through new European Union Directives.[11] After the setting of variable price caps in 1989, regional differentiation in prices increased rapidly in the early 1990s (see the Appendix for a detailed analysis). In the South-West region, which experienced the highest overall rate of increase, a higher original price cap was exacerbated by a mid-period increase. Within the 'basket' of tariffs, the rise in domestic prices was additionally increased by the economic regulator's decision to 'rebalance' domestic and non-domestic, and metered and non-metered customers, thereby reducing cross subsidies. Price increases of up to 20% per year coincided with the boom and bust economic period of 1989–91 while water prices rose by up to twice that of other regions in the country.

The Liberal Democrats (for whom the South-West was a key region of support) successfully tapped the ensuing discontent in the run-up to the 1992 election, causing considerable political anxiety for the incumbent Conservative government. Water was regarded as one of the most serious political issues facing the government after its re-election. With new legislative burdens forthcoming at the EU level, the geographical variation of water prices drew the personal attention of the Secretary of State and the Prime Minister. An *ad hoc* Water Group was assembled as a Cabinet Office official sub-committee, examining water prices and their relationship to EU obligations, lobbying other member states to support the delay in implementation of the Urban Waste Water Treatment Directive, and proposing measures to mitigate the regional differentiation in water prices.

Ironically, these efforts were frustrated by the government's commitment to retaining intact the structure of the post-privatization regulatory framework.

---

[11] The key Directives were 80/778/EEC (Water Quality), 91/271/EEC (Beaches and Bathing-Water Quality), and the then-forthcoming 91/676/EEC (Urban Waste Water Treatment).

Direct subsidies to private companies were unthinkable. Pressure on water companies to provide substantial rebates to customers in affected regions would affect company profitability, shareholder, and City confidence, and undermine the government's ongoing privatization initiatives. Equalization of water charges within water company regions was excluded by the regulatory imperative of fewer cross-subsidies and greater cost-reflectivity of prices. Greater support for low-income consumers through the social security system was precluded by the method by which the Department of Social Services calculated benefit payments; the notional value of utility costs embedded in benefit payments is set nationally (although housing benefit varies geographically). In spite of searching for responses, the government was unable to act. This predicament had been foreseen by Littlechild, the architect of the economic regulatory framework, who argued prior to privatization that setting variable X values would not only impose a high regulatory burden but also be prey to the dangers of inconsistency and alleged unfairness (Littlechild 1986), leading 'customers [to] question on what basis the changes in prices and service levels vary from one [region] to another' (Littlechild 1986: 7). Littlechild went on to anticipate the government's inability to respond: 'if... one [region's] prices are allowed to rise faster than others, the Minister will have somehow to reassure customers that privatisation nonetheless makes them better off than they otherwise would have been' (Littlechild 1986: 7); this proved to be a politically risky strategy.

As water price rises levelled-off in the mid-1990s, and the economy recovered, the issue died down but was then re-ignited by the Labour Party, whose critique of the water industry formed a major plank of its election platform in the run-up to the 1997 election. Following the election, one of the Labour Government's first initiatives was to call a water 'Summit', bringing together water companies and regulators in May of 1997 to announce a ten-point plan[12] intended to reduce some of what it viewed as the privatized industry's worst excesses. Announcing the 'Windfall Tax'[13] on the privatized utilities later that year fulfilled another key Labour election promise, to claw back excessive profits made in the first years after privatization. The promised reviews of water charging and abstraction licensing were initiated under the auspices of the Department of the Environment, Transport and Regions (DETR 1998*a*; 1998*b*; 1998*c*; 1999*a*; 1999*b*), as was a broader review of utility regulation headed by the Department of Trade and Industry (DTI 1998a, 1998b). Due in part to the changed political climate, Ofwat has hardened its stance considerably; significant price limit

---

[12] The points included: mandatory leakage targets; placing water companies under a statutory duty to conserve water in carrying out their functions; legally binding amendments to water company licences requiring compensation to customers affected by drought-related restrictions; and announcements that the government would conduct a broad review of water-charging and metering policies and the abstraction licensing system.

[13] The proceeds of the one-off tax were used to finance the government's 'Welfare-to-Work' programme. The amount paid per utility company was based on the profits made by it in the first four years in the private sector. The tax raised approximately £5bn. in total, of which just over 30% was paid by the water sector.

reductions were announced at the most recent Periodic Review (Ofwat 2000*a*).

The result of these reviews of water policy has been a progressive re-regulation of the entitlements of domestic consumers. The Labour government's stance on social justice issues within the water sector differs distinctly from its predecessor. Bound by their political commitment to privatization and to the perception of its success, successive Conservative governments could not implement significant changes to the regulatory system, even where they wished to do so. With no commitment to the nature of the privatized system as it stood, the Labour government has been far freer to innovate than its predecessor. The Government has also explicitly re-politicized water policy-making, noting in its guidance on water charging that: 'ministers are better placed than an independent economic regulator to consider the acceptability of social impacts on consumers' (DETR 2000*c*, para. 2.18), and stating in its guidance on the designation of 'water scarcity' areas where metering can be more aggressively pursued that: 'this is finally a political judgement, best reserved for the Secretary of State, though acting with the advice of the [Environment] Agency and the Director [of Water Services]' (DETR 2000*c*, para. 5.15).

The changes are significant. Under the provisions of the Water Industry Act (1999), disconnection of domestic water consumers, and other non-private sector users (schools, children's and residential care homes, hospitals) is prohibited, as is the use of limiting devices (for example trickle valves) in the case of non-payment. Company charging schemes will now have to be approved by the regulators, and ministers will be able to give statutory guidance to the economic regulator on charging, and to make provisions for protection for vulnerable groups.[14] Households on low incomes or from vulnerable groups will have alternative charging options made available (DETR 2000*c*; Ofwat 2000*a*). Consumers have the right to optional metering, provided free of charge, and the right to revert to an unmeasured tariff if they so choose.[15] Only 'non-essential' uses such as the use of garden sprinklers or filling of swimming pools are subject to mandatory metering. Rateable values, although outdated, will continue to be used as the basis of water charges for those consumers who choose not to have a meter, although all new homes will be metered.

---

[14] A person is a member of a vulnerable group if s/he 'receives one of a range of six specific benefits and either is entitled to receive child benefit for three or more dependent children who reside in the premises; or is diagnosed, or has a child who is diagnosed as suffering from one of five specified medical conditions, for which they are receiving treatment, and which causes them to use a significant additional volume of water' (Ofwat 2000*a*: 20).

[15] Ofwat released optional metering guidelines in 1992, under which customers could opt to have a meter (Ofwat 1992*c*). Once installed, the meter could not be removed. In contrast, the Labour government's 'optional metering' referred to a policy under which consumers, once having had a meter installed, could opt to revert back to rateable value. Under the Water Industry Act 1999, all unmeasured domestic water customers may ask for a free meter to be installed, with the option of reverting back to an unmeasured charge within 12 months. 'Free' metering (i.e. the company paying the cost of the installation) is required under the Water Industry Act 1999 and was not official Ofwat policy before this time (Ofwat, personal communication). For new homes, and those substantially altered since 1990, metering is mandatory.

The retreat from the principle of economic equity

During the late 1990s, an active debate was initiated about the social policy implications of water charging; the government's ensuing decision to address unacceptable distributive outcomes *within* the system of water regulation rather than through the benefits system stands in distinct contrast to the policies of the first half of the decade. The debate over optional metering deepened the rift between consumer advocates (who favoured low prices), and environmental groups (who argued for higher prices in order to allow for an increase in environmental expenditure). The latter favoured metering, arguing that meters, when implemented with specific (seasonally, temporally, or volumetrically variable) tariffs, will encourage conservation; the former argued that metering will impact negatively, and most severely, on low-income consumers and vulnerable groups. The economic regulator sided with the environmental regulator, supporting metering, arguing that meters enable greater choice by providing customers with information enabling them to control bill levels.

Despite the support for metering, and the underlying principle of economic equity, from both the environmental and economic regulators of the industry, the Secretary of State has introduced regulations designed to protect consumers who face severe hardship 'when they are using large amounts of water for essential purposes and pay on a measured basis' (DETR 2000c, para. 1.5); these customers can opt for a fixed charge equal to the average household charge in their region. This may represent significant savings for some consumers, much greater than the average reduction of £3.67 for the yearly average metered household water and sewerage bill now available to water service company customers through 'social tariffs' (calculated from Ofwat 2000a, table 3). DETR calculated the cost to non-vulnerable consumers of its provisions at less than £1 per household per year across the industry, arguing that this level would be acceptable as it is not a 'disproportionate' increase in charges (DETR 2000c), the cross subsidy being limited by the tight definition of vulnerable groups and limited number of households which may have access.

The government's support for intraregional equalization, and an underlying principle of social equity was made explicit: 'costs should be allocated between different groups of consumers on an equitable basis. Charges . . . should take account of customer's ability to pay, and address the needs of all those on low incomes' (DETR 2000c, para 2.8). The guidance from the Secretary of State for the Environment to Ofwat (DETR 2000c) stated that 'changes to individual bills should not depart significantly from the average of each company, and . . . phasing-in of any large changes should be considered' (Ofwat 2000a: 15). Post-privatization, a degree of intraregional equalization had continued given the persistence of rateable value as a charging mechanism, and the unwillingness to de-average regional charges, given Ofwat's duty to protect rural consumers. The government expressed its opposition to the full application of the principle of 'economic equity' in water charging in arguing that the 'link between water use and cost [established by a meter] is precisely [what] creates the possibility of hardship for customers most in need' (DETR 2000c, para 4.3.1).

In response to water industry and regulatory arguments in favour of extending protection to vulnerable groups through the social security system, or by central financial provisions (such as the cold-weather payments to pensioners to offset fuel costs), the government argued that: 'water is unique: a supply of clean running water is essential to individual hygiene and to public health. The Government wishes to ensure that vulnerable customers do not have to cut down on essential water use, potentially compromising their health or the health of others, because of problems affording charges' (DETR 2000c, para 4.4). The government's stance, it bears noting, was not widely viewed as threatening within the industry. In contrast to its rhetoric, the change in government policy has entailed a return to deliberate, selective cross-subsidies within the water supply system, despite the fact that the industry is now privately rather than publicly owned. However, these cross-subsidies, as noted above, are far more limited in size, and target far more tightly defined groups than those of the 1970s.

These new targeted cross-subsidies may also be offset by the spatial disaggregation of water supply pricing advocated by some companies to facilitate proposed product-market competition in the water supply industry (Consumers' Association 2000). Within each water supply region, customer's bills would be priced according to LRMC for small zones, implying that customers in areas expensive to supply (such as rural areas) will pay significantly more than other customers, who may see their bills drop. Companies would thus have more transparency in revenue streams, which would be more directly related to supply costs; the implications of the resulting reduction in cross subsidy would be a likely increase in the 'cherry-picking' and 'social dumping' seen in other utility sectors. As noted by one of the smaller water-only companies (which would have the most to lose from competition, as the likely outcome of the introduction would be consolidation in the industry):

A safe reliable supply to almost every property is a cornerstone upon which our developed society has been built ... It is important that the cross-subsidies in water which are implicit in regional average charges are not unwound too quickly if at all as this could undermine the Government's socio-economic objectives through incidence effects should these cross subsidies be dismantled. There are also issues related to the director General of Water Services' duties in respect of rural customers who might be one of the first to experience the disbenefits [sic] of competition if de-averaging were to occur because of 'cherry picking' of those customers who are cheaper to supply.[16]

---

[16] Evidence submitted by Bournemouth and West Hampshire plc, in response to DEFRA consultation on competition in the water industry, June 2000.

This strategic 'cherry-picking' is aided by telematics technology, newly introduced to the water industry, enabling water companies to satisfy profitable customers while minimizing the costs imposed by unprofitable customers (Guy, Graham *et al.* 1997). A new system at the South West Water customer service centre, for example, functions such that:

When a customer rings, just by giving their name and postcode to the member of staff, allows all account details, including records of past telephone calls, billing dates and payments, even scanned images of letters, to be displayed. This amount of information enables staff to deal with different customers in different ways. The customer who repeatedly defaults with payment will be treated completely differently from one who has only defaulted once. (*Utility Week* 1995)

These developments within the water sector parallel broader trends in the utility sector, within which 'essential infrastructure resources are [being] commodified and . . . differentiated in terms of cost, availability and quality over space and time' (Marvin and Guy 1997: 1). Generally, utility services post-privatization have experienced a social and spatial polarization of access and cost between low- and high-income groups (Graham and Marvin 1995). The 'rebalancing' of tariff structures, also referred to as 'cost-reflective pricing' has also occurred in gas and electricity industries, in which BPUs are much more widespread than in the water industry (Drakeford 1995; 1997). Within the energy supply sector, distinct spatial patterns in consumption of, and access to energy supply services have emerged post-privatization. These trends are less evident in the water industry, where cherry-picking and social-dumping are constrained by the universal service obligation[17] placed upon water companies and the lack of product competition in the domestic market.

Alternative water charging policies: the false efficiency–equity paradox

Benevolence begins where justice ends. (Sidgwick 1874)

What alternative water charging policy frameworks might be developed to take greater account of social equity? One policy response to regional differentiation of water and sewerage bills might be to regionally (rather than nationally) index the notional value of water in income-support payments, as is already done for housing. More broadly, the implications of alterations to the structure of ownership, and management of demand for water in distinction from the supply of water might be considered. Two issues will be addressed

---

[17] Under the provisions of the Water Industry Act (1991), and unlike other utilities, water undertakers have a duty of universal provision within their area of operation; they must 'provide supplies of water to premises in [their] area and . . . make such supplies available to persons who demand them' (Water Act, s. 37 (I)).

here: the distribution of performance gains under a regime of private, monopolistic supply of essential services; and the apparent paradox between equity and efficiency in water regulation.

First, questions of the mechanisms governing production in relation to distribution have been overlooked in the current debate. As discussed in Chapter 4, water supply companies retain monopolies over the domestic market; competition in the domestic market has not been successfully introduced, unlike the energy and telecommunications sectors. Prices have risen well above inflation since privatization, and are projected to continue rising in line with inflation over the next regulatory period. Over the first decade following privatization, companies made profits well in excess of predictions, significantly outperformed the market, and paid dividends to their shareholders well above the average paid to stock market investors (Ofwat 1996*d*). However, there is no explicit mechanism in place for sharing performance gains with consumers. This was an important source of the critiques of the privatized water industry in the 1990s.

As the analysis presented above has indicated, however, the privatized regime is in many respects better for consumers than its nationalized predecessor. A constructive alternative to monopoly, I suggest, is not nationalization, but rather socialization of the firm. Re-regulation might incorporate recognition of the extent to which customers' bills are directed towards asset maintenance, and a corresponding consideration of their implicit status as shareholders.[18] In the case of the water industry, this might entail what John Kay (1996) has termed the 'customer corporation', in which dividends paid to customer-shareholders would be linked to consumer charges, and incentive structures would differ from the present model (for example the goal of managers could be cost minimization rather than profit maximization for a given quality level, with sharing of out-performance between managers and customers). Following the recent Periodic Review, boards of several companies announced that they were examining various strategies of 'mutualization' (mutual trusts, non-profit companies, friendly societies) that would allow the separation of asset ownership (and liability) from operational management of the water supply system (Ofwat 2000*b*; 2000*c*). As detailed in Chapter 7, some of these proposals were withdrawn in response to regulators' opinions that they were not beneficial to consumers (Ofwat 2000*d*), but water companies are continuing to search for possible modes of restructuring, some of which might entail returning a greater degree of control, if not ownership, to consumers, through the restructuring of incentive mechanisms.

---

[18] The argument refers to the fact that, under current accounting protocols, repairs to pipes—asset maintenance—are categorized as operating expenditure. Customers' bills are thus directed in part towards asset maintenance, as well as to operating expenditure. In an ideal-type model of the firm, all asset-maintenance should be counted as capital expenditure, and funded through equity or debt. Where customers are contributing to the asset base, their funds are being used in a manner analogous to that of shareholders, who expect a rate of return for their investment.

Second, the false paradox between efficiency and equity requires deconstruction. As Harvey noted with respect to social justice and urbanization, 'production *is* distribution and efficiency *is* equity in distribution' (Harvey 1973: 15). Water, like energy, is an unusual commodity because it has a derived demand. Water is used directly for drinking, but it is also converted in appliances (dishwashers, washing machines, radiators, toilets) into the desired service (for example clean clothes, clean dishes, heat, sewage disposal). Whenever water is used, the conversion efficiency of appliances will vary, so issues of capital and running cost are involved. Raising water prices to encourage more efficient water use may also have the effect of cutting consumption of water services, as greater water efficiency requires capital expenditure that lower-income families may not be able to afford. The immediate effect of a water price rise will be a drop in consumption, particularly by the poorest families, unless accompanied by the necessary capital expenditure in the home. Politically, the division between consumer advocates and environmentalist groups might be transcended through a more comprehensive consideration of a plan of capital investment to enable greater conversion efficiency for derived demands in the water sector. This relationship has already been recognized in the energy sector, where policies of providing capital investment in energy-efficient appliances in the home—advocated as simultaneously sustainable and equitable—have been implemented to address 'fuel poverty' (Boardman 1991; 1999). Similarly, 'water poverty' policies might provide subsidies to increase water efficiency in homes, targeted in particular at groups for whom water expenditure is above a threshold percentage of income.

Conclusions

The evolution of water charging policies in England and Wales over the past three decades has been underpinned by a shift in the prioritization of equity from social towards economic equity, and from the ability to pay principle towards the benefit principle. Given the incomplete application of policies of equalization, and the incomplete penetration of metering and application of marginal cost pricing, neither principle of equity has been fully applied in practice. The current consensus—that universal metering is theoretically desirable but impractical and expensive—implies that temporal and spatial cross subsidies will continue in the water sector. The emphasis on economic equity has shifted the balance of these cross subsidies, and the balance of the cross subsidies will continue to shift, particularly if current proposals for increased competition in the industry go ahead.

One of the promises of privatization was that private ownership and management would lead to greater efficiency and lower bills for consumers of utility services. Yet, in the water sector, bills rose steadily above inflation following privatization. Increasingly stringent regulation, and direct intervention by the government on numerous occasions in the late 1990s have sought to mitigate some of the most politically unacceptable effects of perceived poor performance, with negative public opinion exacerbated by NGO and media reports of 'fat cat' salaries and high dividends in the context of consumers' bills rising above inflation since privatization. The reductions in price limits announced in the 1999 Periodic Review were widely viewed as a response to negative public opinion and resulting political pressure on the economic regulator. This resulted in an average reduction in bills for household customers of £30 (DETR 2000c). Share prices fell dramatically after the announcement. The implications of these developments for the long-term future of the water supply industry are considered in the following chapter.

Appendix to Chapter 6: Assessment of some distributive aspects of water charging

This analysis of increases in interregional variation of water charges since 1989 was conducted using data published by the water industry's trade body, Water UK, in its annual *Waterfacts* (2000) (the water service company's trade body, known as the 'Water Services Association' published the 1989–98 issues). The data selected were for unmeasured (i.e. unmetered) household charges; just over 80% of households were charged on an unmeasured basis in 2000 (Ofwat 2000*a*).

Average water and sewerage charges per household have risen significantly since 1989 (Tables 6.5–6.7). Some regions have experienced much higher rates of increase than others (Table 6.7). South-West region—with the highest average unmeasured charge in 1999/00—has the highest ratio of the current charge to that at the time of privatization, and Dwr Cymru — with the third highest

Table 6.5. Average unmeasured water and sewerage charges, per household (1999/00)

Region	(£)
South West	390
Anglian	335
Dwr Cymru	300
Wessex	284
Southern	279
Yorkshire	251
North West	249
Northumbrian	243
Severn Trent	231
Thames	208
Weighted average	277

Table 6.6. Ratio of 1999/00 to 1989/90 water and sewerage bills

Water and sewerage company	Ratio
South West	2.67
Northumbrian	2.38
North West	2.26
Southern	2.24
Severn Trent	2.15
Anglian	2.13
Thames	2.08
Yorkshire	2.06
Wessex	2.04
Dwr Cymru	2.00

Not corrected for inflation

Table 6.7. Percentage rate of increase in average household unmeasured water and sewerage charges

	89/90	90/91	91/92	92/93	93/94	94/95	95/96	96/97	97/98	98/99	99/00	average
Anglian	13	13	17	11	9	7	5	7	5	5	2	8
Dwr Cymru	11	12	16	9	12	8	3	0	4	5	5	8
Northumbrian	8	20	16	11	8	10	7	5	4	6	6	9
North West	11	13	15	9	8	8	6	6	9	6	6	9
Severn Trent	10	12	15	10	9	9	4	3	8	7	4	8
Southern	8	13	16	9	6	7	6	8	7	6	7	8
South West	14	13	16	19	17	14	3	4	0	14	6	11
Thames	8	14	14	7	8	7	6	6	5	5	3	8
Wessex	9	10	15	8	10	6	6	4	4	6	5	8
Yorkshire	15	11	13	8	7	8	3	11	2	5	7	8

Not corrected for inflation
Note rapid increase over the period 90/91 – 94/95, due to higher K values, and higher inflation

Table 6.8. Ratio of highest to lowest average regional unmeasured water and sewerage charges

Year	88/89	89/90	91/92	92/93	93/94	94/95	95/96	96/97	97/98	98/99	99/00
Ratio	1.45	1.56	1.58	1.63	1.76	1.89	1.82	1.79	1.69	1.83	1.88

average unmeasured charge in 1999/00—the lowest (Table 6.6). The absolute difference between average regional prices has also increased; the ratio of highest to lowest average regional price for combined water and sewerage bills rose from 1.55 to 1.88 (Table 6.8). But this measure does not capture the degree of interregional variation. Table 6.9 assesses increases or decreases in regional variation of bills through a comparison of the highest and lowest average regional charges. The yearly change in the coefficient of variation of regional charges (Table 6.10) provides a more accurate method of assessing the change in inter-regional variation across all regions.

Table 6.9. Highest and lowest average regional household water/sewerage bill, 1988–2000

	Highest (£)	Lowest (£)	Weighted average	Highest, % difference from average	Lowest, % difference from average
1988/89	135	93	107	26	13
1989/90	156	100	119	31	16
1990/91	177	113	134	32	16
1991/92	204	129	155	32	17
1992/93	228	140	170	34	18
1993/94	266	151	185	34	18
1994/95	304	161	198	53	19
1995/96	312	171	208	54	18
1996/97	325	182	220	49	17
1997/98	324	192	247	36	22
1998/99	368	201	263	40	24
1999/00	390	208	277	41	25

Table 6.10. Standard deviation, average unmeasured regional household charges

	StdDev
1989/90	19.6
1990/91	21.1
1991/92	25.3
1992/93	30.6
1993/94	37.6
1994/95	43.3
1995/96	43.4
1996/97	43.9
1997/98	41.6
1998/99	48.8
1999/2000	50.9

7

The Retreat of the Market? Re-regulation and Water Supply Industry Restructuring

In May 2001, the assets of Welsh Water were sold to a self-styled 'people's company' named Glas Cymru (Blue/Green Wales). Glas Cymru is a not-for-profit private company limited by guarantee under the Companies Act 1985, with a structure modelled after BUPA—a private, non-profit association which is Britain's largest non-governmental health insurance provider. Glas Cymru was formed for the sole purpose of purchasing the assets of Welsh Water from its American-owned parent company, Western Power. Owned by its members (mostly prominent Welsh citizens), prohibited from diversifying outside of its core water business, and wholly financed by debt, Glas Cymru promised an immediate and sustained reduction in water prices for its customers. Control of Welsh water, it proudly announced, had returned home to the Welsh people.

The news made headlines: 'Has privatization gone full circle'? The query began to seem less far-fetched in the months that followed, as several water companies made restructuring proposals. One English water company made a formal proposal to mutualize; others advertised their assets for sale, and still others proposed a radical refinancing of their core businesses, withdrawing from equity markets on which the public water companies had been floated just over a decade before. As detailed in this chapter, companies have adopted a variety of restructuring strategies. The ongoing restructuring of water companies has challenged fundamental tenets of privatization: that the interests of shareholders can be balanced with those of customers; that water supply is a profitable business for vertically integrated (asset-owning and operating) private companies; that the incentives for efficiency arising from well-regulated equity-financed firms will outweigh their higher cost of capital and deliver lower overall bills for customers.

The first section of the chapter explores the 'regulatory creep' experienced in the industry throughout the 1990s, as the regulatory framework evolved stricter standards, more demanding information requirements, and imposed increasingly stringent 'price caps' on water companies' regulated activities.

The chapter then analyses the ensuing restructuring of the water companies, which have pursued a variety of strategies: diversification, internationalization, mutualization, and refinancing. Mutualization must be understood within the context of these broader patterns of restructuring, as companies have sought out higher-return sectors, in contrast with the capital-intensive, low-return, and increasingly stringently regulated water supply sector.

Re-regulation as 'regulatory creep'

Key to understanding the English water industry is an appreciation of the relationship between the regulatory framework, the conditions of sale of the private companies, and the cash-rich nature of their domestic operations. Water and sewerage companies supply an essential commodity continuously. Throughout the 1990s, English and Welsh water companies had reliable, predictable cash flows. Charges were set in advance for the year, related to the price caps set for a 5-year period. Moreover, the majority of residential consumers (the largest component of water use for most companies) remained un-metered; in 1997/8, for example, 90% of household customers had unmeasured supplies, whereas 80% of industrial customers were metered (Ofwat 1998a). As a result, approximately 65% of turnover for companies throughout the 1990s was based on charges that were not volume-related (WSA 1998). Given that water supply companies continue to operate as regional monopolies for domestic consumers, and as competition to date has applied only to a limited number of industrial consumers, there is little risk to cash flows from competitors in the product market. The water industry (particularly with relatively low levels of gearing in the early 1990s) was relatively cash-rich and secure, with a licensed monopoly over a large stream of revenues (World Bank/IFC 1991). Water managers are well aware of this situation; as one industry newsletter noted:

The consumers have a captive . . . requirement for water and sewerage services and no one else can supply them. The cash generative capacities over many years are vast while long-term capital and debt are relatively cheap. (*Water Briefing* 1995: 10)

Although water companies in the decade following privatization were blessed with relatively secure revenue streams, managers became increasingly aware that future returns to shareholders were by no means guaranteed; on the contrary, the stability of future returns was highly exposed to the risk of a tightening of economic regulatory framework (McGuinness and Thomas 1997: 328). Indeed, 'regulatory creep' was sustained in the water sector throughout the 1990s.

This 'regulatory creep' was at odds with the expectations of the architects of the regulatory framework. The creator of the price cap regulatory regime anticipated that 'light touch' regulation would imply non-intervention in company management: 'once X has been chosen, the regulatory authority does not need to approve price changes nor vet the company's investment programme' (Littlechild 1986: 27). In its original form, then, price cap regulation was intended to entail a light regulatory burden, with relatively small information requirements; companies' price limits were to be set by the regulator once every 10 years, and the opportunity to retain profits within the price limits, backed up by pressure from shareholders and the City, would provide the incentive for efficiency gains. In theory, the key features of the British model are: little regulatory interference; a relatively long time-frame between regulatory interventions; the capping of prices, rather than dividends; and the creation of a system of indicators, or 'yardsticks' which allows simulated competition amongst companies, thus providing incentives to increase efficiency via 'comparative' competition in which the regulator acts as a proxy for the market. Efficiency incentives arise because companies are allowed to retain any profit made after price caps have been set, with the profit motive ensured by 'City' scrutiny; comparative competition enables, in theory, realistic price caps to be set.

Only a few years into privatization, however, key variables upon which the forecasts underlying the price limits had originally been based had changed substantially; the regulator decided that an 'interim' price review should be held, in order to readjust price limits. The decision to carry out Periodic Reviews at 5-year intervals was in part a response to the difficulty of accurate forecasting over even a relatively short period of time. Actual input costs in the period 1990 to 1995 were significantly lower than forecast, resulting in profits above expected levels (Saal and Parker 2001). Price reviews, originally intended to take place once every 10 years, are now conducted every 5 years. The resulting shortening of the review period, combined with instances of regulatory pressure in intervening years on companies not to take up full price caps, has meant that a key incentive underlying the original model of price cap regulation—the ability to retain profits arising from outperformance after the assignment of price caps—has been weakened.

The weakening of the incentive structure underlying the economic regulatory framework is in part also due to the decision in 1989 to depart from the original design of the price cap regulatory framework. Although the architect of the RPI − X system originally recommended that a uniform X value be set uniformly, based on average costs across the industry as a whole (Littlechild 1986), the government chose to set individual X values for each company, in part due to concern over the financial weakness of some companies and the risks of failure at flotation. The practice of setting individual X values has been continued by Ofwat; each company has a different price cap, although the component that relates to efficiency is set nationally. Uniform X values were thought to be preferable because they would facilitate the calculations of the 'industry yardstick' for revising X, avoid accusations of discrimination, minimize the burden of licence negotiations, and avoid the introduction of disincentives (Littlechild 1986). This last point is particularly important given the trade-off between efficiency incentives and savings, as Littlechild noted: 'regulation of prices and profits discourages the search for efficiency by the regulated company insofar as any cost reductions have to be passed onto its customers' (Littlechild 1986: 8).

As Littlechild anticipated in his report on the original structure of the regulatory regime, regulatory creep has imposed a high burden on regulators and thereby on companies (Littlechild 1986). This has arisen not only because of the setting of variable price caps, but also because the economic regulator has been drawn into making implicit decisions on the acceptable real rates of return on capital employed in order to arrive at price limit determinations. Rather than being an end-point of regulation, price caps have become, to some degree, a means of regulating rates of return, via intense scrutiny of and negotiation over the 'true' cost of capital for companies, heralding pronouncements of the 'death' of price cap regulation by regulatory economists.

The economic regulator's focus on rates of return was justified, in part, by concern over consistently high levels of company profits and dividends throughout the 1990s, during which returns to shareholders significantly exceeded the cost of capital, and were significantly higher than those in other countries' utility sectors (Helm 1994; Defeuilley 1998). The regulator's concerns stemmed from the divergence between expected and actual rates of return on regulatory assets earned by water companies since privatization. The expected rate of return was 7% (before financing and corporate taxes); but the water industry's average rates of return had not dropped below 10% since privatization, and were in many instances significantly higher than expected (Miller-Bakewell 1998: 2).

This discrepancy arose, in part, because of information asymmetries between the regulator and water companies. Another contributing factor was the impact of price-cap regulation, which worked well for operating expenditure (creating incentives for efficiency gains between periodic reviews) but less well for capital expenditure. As discussed in Chapter 5, the link between K values and forecast capital expenditure which ensured stable returns for companies created, it was realized, an incentive to inflate investment programmes during the periodic review negotiation process (a variation of the Averch–Johnson effect identified with respect to rate-of-return regulation) (Averch and Johnson, 1962; Helm and Yarrow, 1988). Regulatory scrutiny also increased, therefore, in order to monitor for significant discrepancies between companies' forecast and actual capital-investment programmes. In addition, the difficulty (and cost) of doing meaningful comparisons between companies—whether on environmental or economic criteria—was significantly underestimated by the architects of the regulatory framework (see, for example, Hopkinson, James and Sammut 2000).

As a result, the economic regulator has found it necessary to steadily increase the degree and breadth of scrutiny of companies' activity. Yearly reviews of performance became more exacting. Periodic Reviews became more information-intensive, and more costly; one industry analyst estimated the total cost (to regulators, and to the water companies, who must cover much of the costs of audits and information gathering) of each Periodic Review at somewhere between £30 and £50m. (David Lloyd Owen, personal communication). Water companies are subject to greater disclosure requirements than pre-privatization, and accounting practices have been harmonized; information flows have increased dramatically and their quality improved, in large part due to the information demands of the economic regulator (Sawkins 1995).

Alongside the informal 'creep' of the regulatory framework, the formal regulatory framework has evolved over the 1990s, giving regulators new legal powers and placing new legal obligations on companies. Under the provisions of the Competition Act (1998), for example, regulators have been provided with more potent means for forcing companies with high rates of return to cut their prices between Periodic Reviews, as the Office of Fair Trading has indicated 'profits consistently exceeding a company's cost of capital may be taken to indicate that prices are excessive' (OFT 1999). Ofwat will thus be able 'to use both licence amendments and the Competition Act to control prices that are not competitive and deal with anti-competitive behaviour in areas that are potentially competitive, including prizing open access to monopoly networks' (Parker 2000: 21). The environmental regulator has also gained new powers: to reform the abstraction licensing regime; to require companies to modify environmentally damaging abstractions; and to intervene in company management practices through the imposition of the requirement on companies to formulate drought management contingency plans, subject to approval by the regulator. Simultaneously, political concerns arose from the link between large profits and high prices, which had been rising steeply in real terms since privatization, in contrast to the other privatized utilities. An increase in 'water poverty' with well-documented negative public-health effects (as outlined in Chapter 6) incited consumers' advocacy groups and municipalities to wage public campaigns against the water companies—winning significant court battles against the practices which were seen to impact most severely on 'vulnerable' consumers. The resulting re-regulation of the water supply industry (as outlined in Chapter 6) has imposed additional burdens on water companies. Few, if any, of these reforms were envisaged at the time of privatization.

Financial squeeze

In addition to regulatory constraints, the English and Welsh water companies are subject to significantly greater financial pressure than in 1989. The

Windfall Tax imposed by the government in the July 1997 budget was paid by companies in two instalments, in December 1997 and December 1998, the total amounting to £1.65bn. (1998 prices) which represented over 20% of water industry turnover in the 1997/8 financial year.[1] In addition, tax concessions granted at the time of privatization began declining in the late 1990s. Less important, but not negligible, were the cumulative financial impacts of an increasingly stringent regulatory framework which devolved more of the costs of the core service onto companies. Many of these initiatives were put into place after the election of a Labour government to power in 1997—unseating the Conservatives after nearly two decades in power. Reform of the water sector was a key election platform of the Labour party in the run-up to the 1997 general election. Having won a majority, the Labour government called a 'Water Summit', bringing together companies, government, and regulators, at which it presented a ten-point plan of voluntary and mandatory targets to be met by the water supply industry, requiring companies rather than customers to pay for the installation of water meters, prohibiting disconnection of domestic properties for non-payment, extending the standards schemes under which poor company performance results in customer rebates.

Yet water utilities were simultaneously under pressure from shareholders to sustain high dividends in an environment characterized by what one City financial analyst terms 'Dividend Machismo'—competition between companies to maintain dividends at high levels (Miller-Bakewell 1998). Prior to the announcement of the 1994 Periodic Review, the consensus amongst City analysts was 'buy: 2% real dividend growth pa to 2000'. Following the Review, this position hardened, with analysts recommending the sale of companies, with lower predicted rates of growth and 'buy for income'. Many water companies initiated share buy-backs, in large part in response to City demands to return value to shareholders, often couched in language referring to the need to increase gearing as a means of 'getting rid of financial slack'. As Miller-Bakewell notes, 'sustained premium dividend growth has been the most significant single factor behind the returns—an average of 20 per cent per annum over the past five years—achieved by investors. Indeed, in a UK equity market context, these have been the exception' (Miller-Bakewell 1998: 5). A factor in these high returns has been the erosion of dividend cover.[2]

---

[1] The stated purpose of the Windfall Tax was to claw back gains (or 'windfalls') made by shareholders of privatized companies subject to economic regulation. The argument was made by the government that shareholders received unexpectedly (and presumably undeserved) high returns, because the nationalized assets were sold at a discount, and because of lax economic regulation following privatization. The amount of the levy imposed on each company took into account the difference between the asset value at the time of privatization and a 'more realistic' asset value taking into account profits in the first four years since the privatization of each company. The total amount raised from the Windfall Levy was £5.2bn., over 30% of which was paid by the privatized water companies (Bond, Klemm, and Simpson 2001). Proportional to turnover, the Windfall Tax weighed disproportionately heavily on the water supply companies, as compared to the gas and electricity companies.

[2] Dividend cover is conventionally defined as the ratio of total profits of a business to its dividend payments.

Concerns about water companies' financial health grew in the late 1990s, as industry analysts began warning of a likely reduction of price caps at the next Periodic Review, and consequent reduction in dividend growth. This predicted price-cap reduction had been foreseen since the mid-1990s, given that the fact that 'returns to shareholders [had] greatly exceeded the cost of capital, and exceeded those in other countries' utility sectors' (Helm 1994: 29). Accordingly, Ofwat began warning as early as 1997 that price-cap cuts would be imposed in 2001 (Ofwat 1997e). In the view of the Director General of Water Services, future prices for water and sewerage should fall in order to 'return gains from out-performance—arising both from operating and capital efficiencies—to customers' (Ofwat 1997g). The underlying logic was twofold: cutting price caps would return efficiency gains to shareholders through eliminating 'the excess returns on regulatory assets which have been earned' (Miller-Bakewell 1998: 3) and bring companies in line with 'expected' (and politically acceptable) rates of return.[3]

The 1999 Price Review

In line with predictions, the price limits announced in 1999 for the year 2000–1 reduced bills by an average of 12.4% in real terms, with broadly stable prices until 2005 (Table 7.1). Following the Periodic Review, the regulator confirmed that the target pre-tax weighted-average cost of capital employed in calculating companies' K values was approximately 6.5% (less than 4.75% post-tax, on average) (Ofwat 2002c). Indeed, the average pre-tax return across the sector was 6.6% in 2001/2, substantially lower than the 9.3% return in 1999/2000 (Ofwat 2002c).

In announcing the price cut, the regulator justified his decision in noting that 'the existing ... model has led to greater efficiency of the water companies ... the 1999 price review transferred the benefit of this increased efficiency to customers' (Ofwat 2000d: 4). The reduction in prices also reflected a reduction in the size of the investment programme, as despite strong lobbying by the Environment Agency, Ofwat rejected most of the companies' proposals for increased 'environmental spend', citing the need to reduce the cost burden on consumers. This reduction in price caps is highly significant for the water companies given that the demand for water and sewerage services has been fairly stagnant since 1989. Although revenues have risen, this is largely attributable to the increase in prices allowed by the regulator in order to enable companies to meet capital investment programme targets, most importantly those requirements imposed by European Union legislation. Because demand is

---

[3] Ofwat made this clear in its final determinations report (Ofwat 1999c), in which figure 2 (p. 25) provides three breakdowns of the proportion of consumer prices made up of operating costs, operating profit, and capital charges, with the regulator intervening to reduce operating profit, and hence the return on capital to both lenders and investors.

Table 7.1. Change in effective charging limits following the 1999 Periodic Review

Company	K + RPI + U	
	1999-2000	2000-2001
Anglian	4.5	-8.6
(Hartlepool)	6.3	
Dwr Cymru	3.5	-9.1
North-west	5.5	-7.9
Northumbrian	3.8	-18.0
Severn Trent	3.7	-12.7
South-west	4.0	-10.8
Southern	10.0	-11.6
Thames	3.5	-10.3
Wessex	4.5	-10.6
Yorkshire	5.9	-13.1
Water and sewerage companies	**4.7**	**-10.9**
Bournemouth & W. Hampshire	2.5	-1.6
Bristol	4.0	-8.6
Cambridge	1.2	-12.9
Cholderton	3.5	-6.6
Dee Valley	1.3	-9.2
Essex & Suffolk	5.5	-12.4
Folkestone and Dover	2.5	1.4
Mid-Kent	4.0	-18.3
North Surrey	7.2	-13.6
Portsmouth	1.8	-1.6
South-East	2.0	-14.7
(Mid southern)	2.0	
South Staffordshire	3.5	-1.3
Sutton & East Surrey	-1.7	-15.6
Tendring Hundred	3.5	-5.5
Three Valleys	5.5	-13.8
York	3.3	-7.6
Water only companies	**3.7**	**-11.0**
Industry Average	**4.6**	**-10.9**

Sources: Ofwat 2000a

stagnant, the regulated utility business is highly dependent upon the price-setting regime for revenue growth (Shaoul 2000*b*).

The industry's vulnerability to price-limit reductions was exacerbated by the drop in dividend cover in many companies in the late 1990s. As the economic regulator noted in his June 1997 comments on the forthcoming Periodic Review: 'The Director has consistently set out his view . . . that real dividend growth of between 0–2 per cent p.a. is consistent with the cost of equity for water companies . . . Dividends in the first five years [following privatization] rose in line with higher profits, but more recently there have been significantly higher dividends paid which have been financed from higher borrowings rather than profits' (Ofwat 1997*g*: 43). Sustained dividend growth, in other words, was sustained to some degree by an erosion in dividend cover, and entailed a weakened position for shareholders in the event of a drop in revenue streams.

A drop in share prices following the announcement of the final Periodic Review determinations in mid-1999 was thus anticipated by some industry analysts, although the level of share-price decline was in most cases underestimated; in the event, share prices fell roughly 50% across the water industry (Figure 7.1). In some cases, share prices have continued to trade well below their regulatory asset value (although given that many regulated utilities have been absorbed into larger corporate structures, separating out the regulated element from the diversified structure is difficult). These developments reflect the exposed position of shareholders to the combination of rising financing and taxation costs, and the price cap cuts, implying as a drop in revenues and profits.

The reduction in price limits at the 1999 Review had an economic logic (delivering 'efficiency' gains to consumers, which had up until 1999 been delivered to shareholders), and a political inevitability, given public pressure to

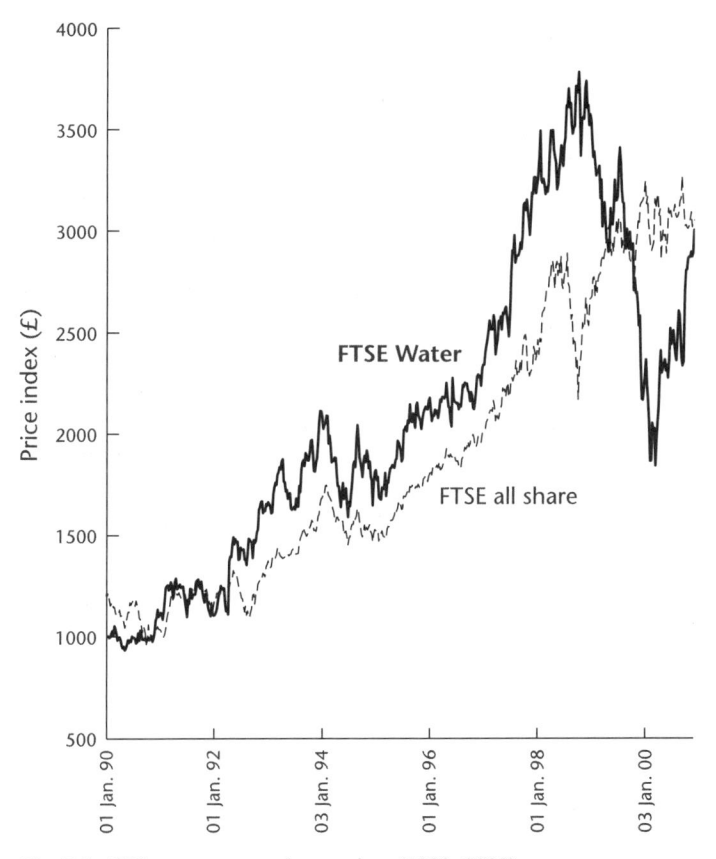

Fig. 7.1. UK water sector share prices (1989–2002).

reduce prices in the industry, after a decade of increasing prices and high profits, during which water companies consistently outperformed the stock market. Failure to reduce the price caps would have further deepened the crisis of political legitimacy of the water industry. Yet the price cap reductions may threaten the long-term financial viability of some of the private water companies, given the wariness of equity markets. The sole means of external finance to which water companies are likely to have recourse in the future is debt financing.

That this should be the case is partly the result of the highly capital-intensive nature of the English and Welsh model of private water provision, in which companies own assets as well as provide operations and maintenance services to the network (Shaoul 2000a; 2000b).[4] In water, approximately two-thirds of the total capital expenditure costs related to distribution; in contrast, distribution and transmission network costs account for closer to two-fifths of total costs in gas, and one-third in electricity (Consumer's Association 2000; WSA 2000). In other words, in contrast to other utility network industries, a high proportion of total costs for the English and Welsh water industry relates to the network rather than to the supply of services. If water prices are held at socially and politically acceptable levels, this capital-intensiveness implies lower income-generation opportunities (Shaoul 2000a; 2000b), which poses difficulties for privatized companies which must cover not only operating costs and taxation, but also dividends and higher interest payments.

This situation is exacerbated by the stagnant demand that characterizes this mature industry, with only very small room for expansion in the regulated domestic market. Given stagnant demand and a legally defined upper bound to the size of the market (i.e. the licensed supply boundary),[5] water companies have little room for growth in the regulated business. Revenue growth is thus to a larger degree than in other networked utility industries dependent on increases in prices. Should prices cease to increase, or drop, revenues will in most cases cease to grow, given that dramatic technological changes in the sector are to some extent constrained (for example by the Building Code, which requires a certain minimum volume of water to be used in some household devices, thereby inhibiting water conservation by users). Given that major technological changes are not imminent, and that labour costs, as a proportion of total costs, are relatively small, efficiency gains are likely to be small. The cut in price caps announced in 1999 thus implies a drop in revenues for most water companies, and a consequent drop in profits, given that a significant proportion of water companies' profits were supported by the trend for regulated output prices to outstrip input costs over the first decade following privatization (Saal and Parker 2001).

[4] The following analysis applies only to asset-owning utilities, not to other types of water providers (such as operations and maintenance services companies, like the French companies Ondeo and Vivendi).

[5] Excluding inset appointments, which are insignificant in terms of total revenues.

Vagabond capital? Multi-utilities, global utilities, and virtual utilities

Given the increasing financial 'squeeze' of the late 1990s, many water companies anticipated greater difficulties sourcing finance to meet future investment requirements. Opportunities for new sources of revenue within the regulated business are limited: demand is likely to remain stagnant as the industry is mature; water utilities cannot increase market share (except a negligible proportion via inset appointments); and future efficiency gains are believed by company managers to be limited (Vass 2002).

One strategy for companies entails increasing productive efficiency; perhaps predictably, several companies announced job cuts immediately following the announcement of the reductions in price caps. Indeed, post-privatization, all water companies have commercialized employment and management relations as a source of productive efficiency gains (O'Connell-Davidson 1993; 1990; Du Gay and Salaman 1992; Mulholland 2002; Ogden 1997). Cost-centres, performance-linked pay rises, and share ownership schemes have been introduced to incentivize management, for example. The impact of commercialization on labour has been a reduction in the size of the workforce, and also a reduction in the proportion of employees holding 'standard' direct contracts —allowing them full-time, unionized, stable employment with good fringe benefits. The increase in part-time and informal contracts has allowed management, in general, to force concessions from labour—in terms of working hours, working practices, and work intensity (O'Connell-Davidson 1993).

Linked to commercialization of employment relations has been a strategy of outsourcing. Under the current regulatory regime, where prices are capped and little if any expansion of the consumer base is possible, one strategy for companies operating close to the efficiency frontier—particularly on OPEX— is to focus on reducing the cost of capital as compared to the current regulatory benchmark, through transferring risks 'upstream' by contracting out (Vass 2002). Outsourcing has occurred in two ways: spinning off companies from the regulated businesses which, staffed by former employees, render the same services on a consultancy basis; or subcontracting non-core (for example cleaning, computer programming), and even core functions (for example pipe maintenance). These strategies have permitted companies to continue the trend of reducing the core workforce initiated in the early 1980s.

The difficulty of increasing profits in a mature, regulated, low-growth industry is also an incentive for diversification. Diversification of the water business was initiated by all of the privatized water companies in the years following privatization. Theories of corporate diversification tend to focus on three drivers: the opportunity to increase market power; managers' self-interest; and the opportunities for developing sustainable competitive advantage (McGuinness and Thomas 1997). In an industry as highly regulated as the domestic water business in England and Wales, however, the role of regulation is another key factor (Bauer 1995). Management 'justify diversification in terms of a need to secure alternative sources of income and profit' (McGuinness and Thomas 1997: 328), in light of the gradual tightening of the regulatory framework.

Given perceived low-growth opportunities in the domestic market, and the difficulty of sourcing finance to meet future investment requirements (which had been anticipated well before the 1999 price limits were announced), water companies in England and Wales have opted to intensify three restructuring strategies. *Diversification* entails remaining vertically integrated and growing through takeovers and mergers and/or expanding into non-regulated businesses and other utility sectors. *Vertical de-integration* implies unbundling activities and in some cases hiving off the regulated business to concentrate on non-regulated activities. *Internationalization* involves becoming an international water business, which may or may not entail diversification and vertical de-integration.

Diversification

At privatization, the RWAs became publicly limited companies (plcs). A group plc structure was adopted by all WaSCs and some WoCs (Figure 7.2). The core business of water and sewerage services was transferred to a subsidiary company acting under a licence (formally known as the Instrument of Appointment) granted by the Department of Environment, Transport and Regions. The boundaries of water supply areas were unchanged by the transfer to private ownership. In their existing regions of operation, the WaSCs were appointed as vertically integrated regional monopolies, providing the entire cycle of services to their customers, from extraction of raw water, delivery of processed water, and collection, treatment, and discharge of waste water. The (smaller) private water supply companies were appointed to provide water services to their previous customer base.

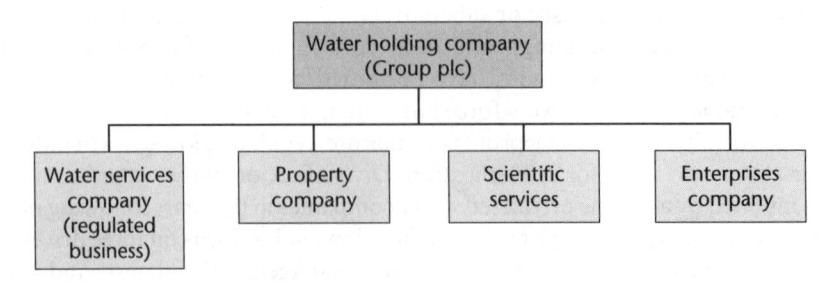

```
                    Water holding company
                         (Group plc)
         ┌──────────────────┼──────────────────┐
         │                  │                  │
  Water services        Property         Scientific        Enterprises
     company            company           services           company
    (regulated
     business)
```

Fig. 7.2. Schematic group plc corporate structure.

Since 1989, all 10 WaSCs have diversified operations (Table 7.2). Attracted by the opportunities for increasing market power and developing a strategic competitive advantage, some utilities have pursued a strategy of 'horizontal integration', creating multi-utilities spanning the electricity, gas, water and telecommunications sectors (e.g. United Utilities, and (formerly) Scottish Power). Others have pursued joint ventures and 'pure' diversification into unrelated sectors. The water companies have diversified into businesses as diverse as land and property development, shopping malls, process engineering, and the hotel trade (OXERA 1994*b*; 1995; 1996). In general, water companies have followed four strategies of diversification (Ofwat 1996*b*). All have transferred some areas of the core business (such as research services) into separate companies while continuing to trade largely with that company for these services. The companies have also segregated non-core activities into one

Table 7.2. Takeovers and diversification of the water and sewerage companies, 1989–2001

Company	Status	Domicile of parent	Major activities
Anglian	Merged with Morrison Construction & Property; now trading as AWG	England	Water, environmental services, construction
Dwr Cymru (Welsh Water)	Merged with South Wales Electricity to form Hyder, acquired by Western Power; assets later sold to non-profit management company	USA	Multi-utility (electricity, gas, water); now water-only
North West Water	Merged with Norweb to form United Utilities	England	Multi-utility (water, electricity, telecommunications)
Northumbrian Water	Acquired by Lyonnaise des Eaux (ONDEO)	France	Water, environmental services
Severn Trent Water	Merged with Biffa Waste Services; remains independently listed	England	Water, waste management
Southern Water	Acquired by Scottish Power (Currently subjct to Vivendi takeover bid)	Scotland	Multi-utility (water, electricity)
South West Water	Now trading as Pennon Group	England	Water, waste management
Thames Water	Acquired by RWE	Germany	Water, environmental services, construction
Wessex Water	Acquired by Enron subsidiary Azurix, then YTL Holdings of Malaysia	Malaysia	Multi-utility (water, electricity)
Yorkshire Water	Now trading as Kelda Group	England	Water, environmental services

Source: water company annual reports, various years

or more group subsidiaries and attempted to sell these services to third parties as well as to the parent company. Most have acquired new subsidiaries and undertaken non-core investments, which may not involve water and sewerage activities; and most have engaged in non-core activities, such as, for example, joint ventures in public water supply systems outside their monopoly licence area (Ofwat 1996*b*).

Companies have followed distinct strategies. Severn Trent, for example, has focused on the environmental services market. Anglian (AWG) has focused on 'private sector participation' (PSP) contracts in asset construction and management (Global Water Intelligence 2001). Kelda diversified into waste management and IT services. Some diversification has taken place within the UK. The reorganization of the Scottish water industry has introduced a new requirement for all major new capital projects to be carried out as competitively tendered BOOs (Franceys 1997); competing for Scottish water and sewerage contracts through non-core subsidiaries has been one state-directed impetus for water company diversification. The compulsory competitive tendering requirements placed by national government upon local authorities for various services (Patterson and Pinch 1995) have also enlarged the potential market for water and sewerage companies, most notably in the waste services sector (Curwen 1994). Given the synergies with skills built up in the core business, waste management has been a common feature of diversification by many water companies; dealing with industrial trade effluent on a 'semi-commercial' basis has been an entry into the sector for many companies.

Diversification is spurred by pressure from the 'City' and capital markets. Only those companies who had already established a secure base of unregulated earnings were perceived, in the first decade following privatization, to be likely to remain good investments. A second incentive for diversification is the threat of takeover, no idle threat given that 5 of the 10 original water and sewerage companies have been taken over, and the number of water-only companies has been reduced from 29 to 13 through mergers and acquisitions (Table 7.3, Figure 7.3). By 2002, foreign companies had taken over 4 WaSCs[6]

---

[6] Approximately half of the smaller water-only companies are now foreign-owned. Of the 10 water and sewerage companies, 4 have successfully been purchased by foreign companies. Northumbrian merged with a water-only company in the same region (North-East Water) when it became part of the French Lyonnaise des Eaux group in 1994, the first major take-over of a private water and sewerage company. Enron, an American-based energy multinational, took over Wessex Water in September 1998 (BBC 1998); after Enron collapsed, Wessex was purchased by YTL Holdings of Malaysia in 2002. German multi-utility RWE purchased Thames Water in 2000, creating the third largest water utility company in the world (after the French companies Vivendi Environnement and Lyonnaise des Eaux/ONDEO). The US-based utility company Western Power purchased Welsh Water, but later sold the core, regulated company as part of the creation of the Glas Cymru mutual. In mid-2002, Vivendi Environnement (already an owner of several, smaller water-only companies) made a bid to buy Southern Water; the deal would have given Vivendi approximately 10% of the English market. At the time of writing, the UK's Competition Commission had concluded that Vivendi's acquisition of Southern Water would be against the public interest because the reduction in the number of comparators would impact negatively on the ability of Ofwat to administer the comparative competition regulatory framework (Fletcher 2003).

Table 7.3. Consolidation of the water industry, 1989–2001

1989 (39 companies)	Intervening years	2001 (23 companies)
Anglian Water Services Ltd	Anglian Water Services Ltd	
Hartlepool Water plc		
Bournemouth & District Water Company		Bournemouth & West Hampshire Water Company
West Hampshire Water Company		
Bristol Water plc		Bristol Water plc
Cambridge Water Company plc		Cambridge Water Company plc
Cholderton & District Water Company		Cholderton & District Water Company
Chester Waterworks Company		Dee Valley Water plc
Wrexham Water plc		
Dwr Cymru (Welsh Water)		Dwr Cymru (Welsh Water)
Folkestone & Dover Water Services Ltd		Folkestone & Dover Water Services Ltd
Mid-Kent Water plc		Mid-Kent Water plc
Northumbrian Water Ltd	North East Water plc	Northumbrian Water Ltd
Newcastle & Gateshead Water Company		
Sunderland & South Shields Water Company		Northumbrian Water Ltd
Essex Water plc	Essex Water plc	Essex & Suffolk Water plc
East Anglian Water Company	Suffolk Water plc	
Portsmouth Water plc	Portsmouth Water plc	South Staffordshire Water plc
Severn Trent Water Ltd	Severn Trent Water Ltd	
East Worcester Water plc		
Eastbourne Water Company	South East Water plc	
Mid Southern Water plc		
Mid-Sussex Water Company		
West Kent Water Company		
South Staffordshire Water plc	South Staffordshire Water plc	
Southern Water Services Ltd		Southern Water Services Ltd
South West Water Ltd		South West Water Ltd
Sutton District Water plc		Sutton & East Surrey Water plc
East Surrey Water plc		
Tendring Hundred Water Services Ltd		Tendring Hundred Water Services Ltd
Thames Water Utilities Ltd		Thames Water Utilities Ltd
Colne Valley Water Company	Three Valleys Water Services plc	
Lee Valley Water Company		
Rickmansworth Water Company		Three Valleys Water Services plc
North Surrey Water Ltd		
North-West Water Ltd		United Utilities Water plc
Wessex Water plc		Wessex Water plc
Yorkshire Water Services Ltd		Yorkshire Water Services Ltd
York Waterworks plc		

Reproduced with kind permission from Page and Swyngedouw 2002

160   *Re-regulating the Water Supply Industry*

Fig. 7.3. Water service and supply companies, England and Wales (2000).

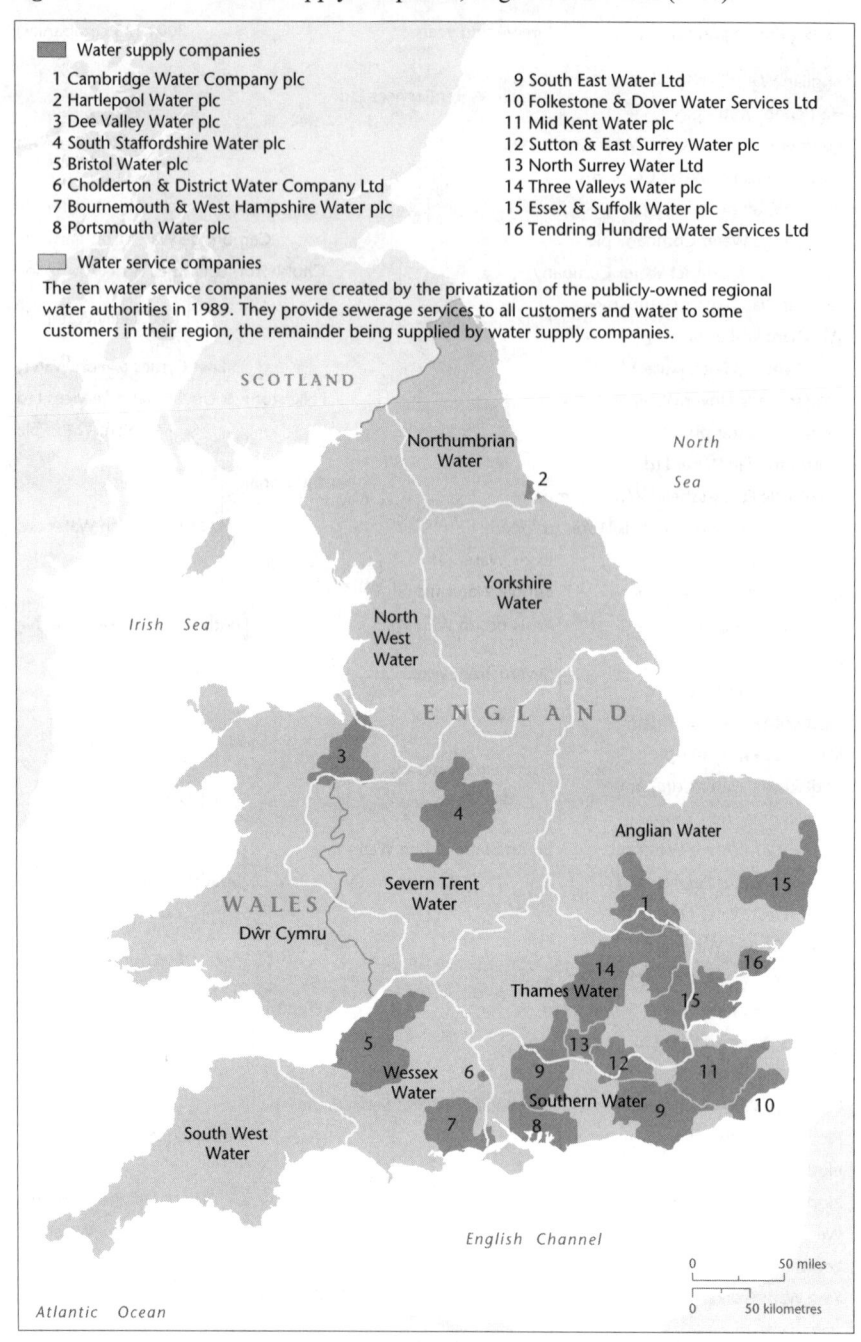

Water supply companies
1 Cambridge Water Company plc
2 Hartlepool Water plc
3 Dee Valley Water plc
4 South Staffordshire Water plc
5 Bristol Water plc
6 Cholderton & District Water Company Ltd
7 Bournemouth & West Hampshire Water plc
8 Portsmouth Water plc
9 South East Water Ltd
10 Folkestone & Dover Water Services Ltd
11 Mid Kent Water plc
12 Sutton & East Surrey Water plc
13 North Surrey Water Ltd
14 Three Valleys Water plc
15 Essex & Suffolk Water plc
16 Tendring Hundred Water Services Ltd

Water service companies

The ten water service companies were created by the privatization of the publicly-owned regional water authorities in 1989. They provide sewerage services to all customers and water to some customers in their region, the remainder being supplied by water supply companies.

and 16 of the original 29 water-only companies. Diversification is thus both a response to, and a driver of consolidation of the industry.[7]

Opportunities for further consolidation of the industry are, however, perceived to be limited: rationalization within the industry is thought to be one means of producing cost-savings for merger counterparts, but mergers have been limited by the regulators' insistence on the need for a sufficient number of independent water companies to ensure 'healthy' comparative competition. Because of this requirement, the economic regulator has made clear his opposition to new mergers between licensed water and sewerage companies (as distinct from the smaller water supply companies).

Financing diversification

The 10 WaSCs spent £1.274bn. on acquisitions by March 1994 (Schofield and Shaoul 1997). Diversification was funded in part from loans, but also in part by earnings from the core, regulated water and sewerage business. Earnings moved out of the utility into the parent company in several ways. First, the regulated utilities, which in the 1990s typically represented 90% of the turnover of the parent group, might pay dividends to their parent company. In the years following privatization, 'dividends paid by the regulated businesses to their parent groups have exceeded those paid by the groups to shareholders' (Ofwat 1995a: 3). This practice was of concern to Ofwat, which argued that dividend payments made by the core, regulated utilities were not 'justified by reference to what would be paid to shareholders in a comparable free-standing plc' (Ofwat 1995a: 3). As Ofwat noted, the payment of excessive dividends 'results in cash resources leaving the core business' (Ofwat 1995a: 27). After dividends, money is retained by the parent plc for its own operating expenditure and business development in non-core sectors. Some dividends are recycled back to the regulated business in the form of interest-bearing debt to make up the shortfall between the amount needed for capital expenditure and cash available after paying dividends to the parent company (Shaoul 2000b: 17). Shaoul (1997) calculated that the industry-wide average for this amount was over 25% of the dividends paid by the utility to the parent company over the first regulatory period.

A second means by which capital moves from the utility to the parent company is via payments for services rendered by other group companies. Non-core companies, in other words, charge more (and in some cases much more) than the market rate for goods or services supplied to the core utility. The possibility that transfer payments amongst the different companies within the plc were excessive sparked an investigation of competitive tendering practices in 1996 (Ofwat 1996e). The investigations 'revealed a number of practices which I

---

[7] Further consolidation of the industry may be limited. Rationalization within the industry is thought to be one means of producing cost-savings for merger counterparts (*Water Briefing* 1996a), but mergers have been limited by the regulators' insistence on the need for enough independent water companies to ensure 'healthy' (if artificial) competition.

do not consider desirable' (Ofwat 1996e: 2), leading consumer advocates to call for separate listing of the plc group companies on the Stock Exchange.[8]

Other methods for moving capital from the utility to the parent company include loans to the parent company, supported by the core company, for funds used to build up non-core business. The allocation of joint costs across the plc when the regulator considers allowable price increases for water and sewerage services also subsidizes non-core activities with core funds:

> The companies will have an incentive to allocate as much of the joint costs, such as management time, into the core business, where cost recovery is guaranteed. This would penalise water consumers and could lead to unfair competition in the non-core markets into which the company has diversified. (Cowan 1994: 122–3)

Water companies have also recycled dividends from the core to the parent company as interest-bearing loans back to the utility, the parent company thereby earning interest from the core company (Shaoul 1997). These movements of capital may be characterized as a subsidy of the non-regulated businesses by utility customers and represent significant sums. The payment of dividends by water companies to parent companies during the first regulatory period was more than £5.6bn., more than the original collective purchase price of the utilities (Shaoul 1997). More subtly, the core domestic business typically represents a solidly profitable subsidiary that lends confidence to investors such as the European Investment Bank, which enables the parent company to secure loans for use in non-core subsidiaries which it could not otherwise obtain. Earnings from the core, regulated business have thus been used to support the diversification activities of the water companies.

Internationalization: global utilities

As early as 1998, British Water (the leading trade body of the British water industry) was predicting a decline in capital investment in the industry after 1999, and arguing that, as a result, 'there is a clear need to develop alternative markets' (British Water 1998: 3). Conventional wisdom post-privatization held that development of alternative high-growth markets requires water companies to move abroad, given that the English markets were saturated, and that the highest growth rates in the industry would be in industrialized countries with low penetration of the private sector (Canada, USA, some Western, and especially Eastern European countries), and to a lesser extent in developing countries.[9]

---

[8] This criticism has been levelled with greater emphasis at French-owned companies. All eight French-owned water companies were found to have paid group charges to their parents that were 'not directly linked to services provided' (Ofwat 1996e).

[9] One study identified 171 projects world-wide with a total worth of over US$35bn. whilst noting that the US market alone, of which approximately 15% is private, has produced at least 25 projects valued at US$3.2bn. in 1995 (Global Water Report 1996).

The move to internationalize operations was initiated before privatization in 1989. Prior to privatization, many of the RWAs engaged in overseas ventures. The 1983 Water Act gave companies the right, for the first time since nationalization, to engage in advice, assistance, and training to overseas clients, but not direct investment. None of these restrictions applied to the consulting engineers and process contractors who engaged in overseas water business; the Department of Trade and Industry (DTI) met regularly with non-RWA water professionals to assist with securing capital projects overseas. By the mid-1980s, the RWAs were included in these discussions, and were also making use of British tied aid, through bidding for overseas water contracts offered by the department of Overseas Development Aid (NEDO 1987). However, as approval had to be sought from the Secretary of State for each venture, who also capped the total value of each RWA's non-core activities, working overseas was not regarded as a major part of water authority business (NEDO 1987).

Nonetheless, well before privatization, the opportunities for private sector involvement overseas were recognized:

There is a large potential market both in the Third World and in developed countries for water goods and services. The World Bank estimates that the total long-term investment required to provide the world's population with clean water and sanitation is £250 billion ... there is potentially a big market and there is scope for increasing Britain's share. (NEDO 1987)

In its eagerness to gain a larger global market share, the UK water industry is not alone. The activities of infrastructure entrepreneurs in 'developing' countries in various sectors, most notably energy and power, have increased substantially over the past decade. With disappearing opportunities for dam-building at home, for example, Nordic and North American dam-building expertise has been vigorously exported, often with substantial assistance from domestic governments, and bilateral and multilateral agencies (Usher 1997). For the English and Welsh water companies, the disappearance of colonial markets was another important factor:

The benefits to many sectors of UK industry from past colonial connections have all but disappeared. In the future, providing management and training facilities for developing countries is one of the keys to maintaining and expanding opportunities for UK industry. If this is not done, other countries will gain market share at our expense. (NEDO 1987)

In light of this recognition, by the mid-1980s, government ministries (most importantly, the Department of Trade and Industry (DTI) and the Overseas Development Agency (ODA)) actively supported the export of water sector expertise and products in anticipation of privatization. Post-privatization, the role of the national state in facilitating water industry exports has increased. The DTI has a full time Export Promoter for Water, seconded from the water industry. In conjunction with the DTI, the Foreign Office and the Water and

Regulation Division of DETR also organize 'outward missions', and support water company's overseas investment activities. In addition to the facilitating role played by country and regional desks, the DTI's Infrastructure and Energy Projects Division has funded 'outward missions' for the water industry to several countries including India, China, the United States, Mexico, and Indonesia, and received 'inward missions' from numerous countries.[10]

The relationship with the Department for International Development (DFID, formerly ODA), the department responsible for British overseas aid, has also remained strong. The DFID runs several programmes aimed at increasing private sector involvement in water development projects (DFID 1998a; 1998b). It supports research into private sector participation; its most recent product is the most comprehensive document on private sector participation in the water and sanitation sector to date, the *Toolkits for Private Sector Participation in Water and Sanitation*. Produced in conjunction with the World Bank, the 'Toolkits' are intended for government officials of middle-income developing countries, informing them of the private sector partnership alternatives (Franceys 1997: 17). In conjunction with WaterAid, the UK water industry's development charity, the DFID sponsors 'Business Partners for Development', an initiative launched by the World Bank that brings together the UK water companies, NGOs, and the DFID to 'maximize the impact' of provision of water and sanitation services to the urban poor. In addition, the DFID conducts research on privatization options for lower-income countries, and runs development programmes for government officials providing 'commercialization and pre-privatization preparation for senior staff responsible for running water and sanitation supplies' (Franceys 1997: 17). The geographical distribution of responsibility between the DFID and the DTI has been further entrenched in handling water privatization. The DFID handles water privatization ventures in lower-income countries (with a focus on sub-Saharan Africa), with aid money and concessionary loans available from agencies like the World Bank, and the DTI handles water privatization in middle-income countries, liaising with 'enabling agencies' such as the International Finance Corporation.[11]

The water industry has joined forces with consulting and processing engineering firms, construction companies, and materials suppliers to form British Water, established in 1993 to act as the 'one-stop shop' in Britain for foreign companies and governments wishing to do business with the UK water sector (British Water 1995). This industry group, representing over one hundred companies, makes clear its continued dependence on government intelligence,

---

[10] Interview, DTI, July 1999.

[11] DTI has an 'A' list of nine countries targeted for water sector activity: Brazil, Canada, Chile, China, South Africa, India, Indonesia, Philippines, and Poland. These countries were selected on the basis of having large urban centres, with low connection rates, an expanding middle class having the ability to pay private sector water rates, a government climate favourable to privatization, and an 'affinity' towards Britain (interview, DTI, July 1999).

contacts, and assistance: 'only through routine contact with [DTI] country desks and the Export Promoters can [we] track the international market opportunities that form the basis of [our] overseas programme . . .' (British Water 1995: 11). British Water regularly holds meetings attended by civil servants from the DTI[12] and the Water Directorate of the DOE to 'determine the optimum ways of aligning the commercial aspirations of British Water members with the government's own commercial policy and to channel the various means of official support for industry into a cohesive policy for the benefit of members' (British Water 1995: 11).

Central government support is particularly solicited as UK water companies are at a disadvantage in exporting their asset sale version of privatization abroad. No other country in the world has followed the UK asset sale route for water privatization, preferring the French concession and *affermage* approach. Indeed, the majority of the international ventures undertaken by the English and Welsh water companies entail the operation and maintenance of infrastructure and/or construction services under contract to governments who own, or will eventually own the assets.

Thames Water goes global

Many water companies have attempted expansion overseas, bidding for concession contracts, or for Build-Own-Transfer contracts. The drive to internationalize began immediately after privatization. This capital was, to a significant degree, not generated through loans; WASCs became involved overseas 'in search of future earnings growth with the cash and loans generated by the core businesses' (Schofield and Shaoul 1997: 13). For most water and sewerage companies, the core utility company provided a secure and profitable base from which to internationalize.

The water companies' early diversification and overseas acquisitions are widely regarded as failures by the City, shareholders, and regulators alike (Dale 1995). Many of their overseas investments have been widely criticized. The parent companies' subsidiaries and acquisitions have not been very profitable: 'in 1994 the core businesses accounted for 85 per cent of sales but 101 per cent of profits of the parent companies, and this trend has continued' (Schofield and Shaoul 1997: 13). In 1995, operating profits for the water industry's non-core businesses represented less than 3% of the total (Martinson 1996*e*). Since the mid-1990s, many water companies sold their poorest-performing companies. Only the larger water utilities have retained a strong commitment to internationalization.

Thames Water, the largest domestic company (in terms of numbers of customers), has developed one of the largest international portfolios of the

---

[12] Specifically, the Projects and Export Policy and Overseas Projects Support Divisions (British Water 1995).

English and Welsh water companies. From 1989, the company pursued a strategy of overseas investment and diversification. These initial acquisitions were based in part on experience gained pre-privatization, as Thames—unlike the other water authorities—had its own marketing organization (NEDO 1987), and offered limited consultancy and training services (Thames Water 1991). Other water authorities worked through British Water International, the former overseas consultancy firm of the UK water authorities, or were contacted directly by clients or consulting engineers on behalf of clients, operating for the most part on a small scale, as consent was required from the Secretary of State prior to working overseas under the Water Act 1983.[13]

As the largest of the water companies, with no debts, cash-rich, and with previous experience of overseas work, Thames Water was well placed to move into the international market at privatization. In the first year of privatization, Thames Water either fully or partly owned 26 companies (Meredith 1992); by 1993, this had risen to over 40 companies, based in 15 countries and trading in over 70 (Thames Water 1993). Germany, the USA, and Southeast Asia were targeted for investment.

TW's early non-core activities did not perform well. Early losses were downplayed; by 1994, however, it was clear that the non-core businesses were not performing as expected. In 1994, the company declared a £33m. exceptional item for losses on non-core activities (Figures 7.2 and 7.3). Losses continued in 1995, and by 1996, with losses of £41m. on a turnover of £92m., the company decided to 'strategically withdraw' from much of its non-core business (Thames Water 1996). In 1995, a number of companies in Australia, South Africa, and the USA were sold; this was followed in 1996 with the sale and closure of others in Germany, Egypt, and the USA, with a non-operating exceptional charge of £95m. (Thames Water 1997). By 1996, TWUL had sold 60% of the businesses acquired in earlier diversification ventures (*Financial Times*, 22 Mar. 1996). Non-core business was not abandoned; rather, a new strategy focused on 'group business in areas more closely associated with the Thames core expertise in management and operation of water and wastewater systems' (*Water Briefing* 1996*b*: 8). Joint ventures and partnerships rather than stand-alone projects became the new emphasis (Garrett 1996).

By the mid-1990s, in contrast to its early internationalization ventures, Thames Water International was refusing contracts for design and construction, entering instead into joint ventures with construction partners where necessary. Priority was gradually given to operation and maintenance and on management contracts, approaches with less revenue risk, than on the more ambitious BOT projects, in which it previously entered as an equity partner.[14] This strategy was reinforced domestically by Ofwat's repeated statements that

---

[13] In practice, consent was rarely given, and the amount of potential profit from overseas work remained very small in relation to turnover of the core business in the UK (NEDO 1987).

[14] Lex Column (1998), *Financial Times,* 'Thames Travails', 26 May, 24.

water companies' generous dividend policies were unsustainable. Ofwat's hints that tough new targets for price cap limits and leakage would be forthcoming in the 1999 Periodic Review (see, for example, Taylor 1998) reaffirmed Thames strategy of diversification. TW new goal was to increase the earnings from its non-core businesses, with a target of 10% of profits from overseas business by year 2000.[15] By 1997, the company declared a small profit on overseas ventures in Turkey, Australia, Malaysia, Thailand, China, Puerto Rico, and Indonesia. By 1998, it held a US$3bn. billion project portfolio, including the 'largest privately financed water scheme in the world',[16] a BOT in Turkey. By 1998, City analysts judged that, of the 10 water and sewerage companies, 'only Thames Water has so far established the probability of a meaningful profits stream immediately post the impact of the 1999 Periodic Review' (Miller-Bakewell 1998: 38). By 1999, the company had grown to 24 million customers, the majority of which were located outside the Thames basin, and hence not serviced by the core, domestic regulated business. A cautionary note should be sounded about the significance of the international side of the water business; Thames Water International only first showed a profit only in 1996/7, and the turnover and profits of the core business are 10 times that of the international business.[17] Nonetheless, Thames Water's securing of substantial unregulated earnings beyond the 1999 Periodic Review given 6 major international contracts, had made the company one of the strongest UK WaSCs. With 20 operations in 14 countries and 19 offices abroad by the end of the 1990s, Thames had grown to become the fifth largest international water services company in the world. The buyout of Thames Water by the German multinational multi-utility RWE in late 2000—creating the third largest water utility in the world when combined with RWE's water activities—thus came as no surprise to industry observers.

Vertical de-integration: unbundled utilities

In the first years following privatization, diversification and internationalization remained preferred strategies of water companies (Ofwat 1996c). In the absence of competition in the domestic market, the privatized water companies remained vertically integrated regional monopolies, owning and operating the assets used to deliver services, and thus remaining responsible for network maintenance as well as service delivery. Soon after the 1999 Periodic Review,

---

[15] Interview, Thames Water International, September 1998.
[16] Source: http://www.Thames-water.com/international.
[17] In the financial year 2000, Thames' international operations had an operating profit of £43.7 m., on a turnover of £151m., with 2,000 employees; the domestic arm had an operating profit of £487.7m., on a turnover of £1.126bn., with over 5,000 employees (source: http://www.rwe.com/eng/1.8_RWE_Konzern/Konzernstruktur/Multi-Utility/Thames_Water/Thames_Water_Profile.html).

various water companies put forward proposals for the vertical de-integration of the industry—separating ownership from operation. From the perspective of managers of some group companies, unbundling activities along the supply chain—and particularly devolving asset ownership, perceived to carry increasing risk and decreasing profitability—was thought to be a strategy which would improve overall performance.

Vertical de-integration would allow the asset owners to contract out the supply of services on the basis of competitive bids from water operator companies—similar to the system in France, in which the owners of water networks (almost always municipalities) award long-term contracts through a process of competitive tender (albeit marred by corruption in recent years (Hall and Lobina 2001)) to private companies to operate water and sewerage networks. Vertical de-integration would thus represent the most significant structural change to the industry since privatization.

One motivation for restructuring stems from the difficulty experienced by some water companies, particularly those whose shares are trading below regulatory asset value, to source finance for future capital expenditure programmes, due to the pressures of the windfall tax and the price limit reductions imposed in 1999. The capital programme for the 2000–5 period has been reduced in comparison with the 1990s, but at £6.4bn. pounds is still significant (House of Commons Environment Audit Committee 2000: para 188). Restructuring, it is thought, would allow for cheaper financing than the current equity-intensive model, through allowing companies to achieve a lower cost of capital due to a changed risk profile. Simply put, the ownership and provision of assets is viewed as a low-risk activity that can be funded by cheaper, long-term debt; operations and customer service are riskier, and would be appropriately funded by more expensive risk capital.

Some companies, it should be noted, have sought alternative modes of financing without proposing to vertically de-integrate. Sutton & East Surrey, for example, has chosen to increase its gearing (debt/equity ratio) to 75%, thereby refinancing existing debt, and to fund future capital expenditure with an innovative index-linked sterling bond, allowing the company to raise debt at a lower cost of capital (*Utilities Journal* 2001*a*). Yet lack of access to sufficiently cheap financing is not the sole motivation for companies interested in mutualization. The assumption that the return on capital for services companies is more attractive than for asset owners under the more stringent price caps now imposed by the regulator is another motivating factor. While facilities management companies typically have low profit margins, they are able to generate a respectable rate of return on capital employed because they have few assets; the move to restructure, in other words, should be viewed in the same terms as the broader trend in industrialized economies towards service provision over the past three decades (Shaoul 2000*a*). Another explanatory factor is that some City analysts have a vested interest in mutualization, which has been 'spurred on by consultants [in corporate and bond finance] who hope to make considerable sums from inventing and facilitating the necessary arrangements' (Summerton 2001).[18]

In the long-term, however, the declining status of water and sewerage infrastructure may be the key factor in companies' desire to exit the asset ownership side of the business. Little of the asset stock in Britain is less than three decades old; much of it is older than 50 years, and in some cities, older still—in London, one in three kilometres of pipes is more than 100 years old (Perera, Farley *et al.* 1985). The large discrepancies between Ofwat and industry estimates of necessary levels of funding for capital maintenance to maintain infrastructure, much of which may be up for renewal at the same time given large historical variations in asset installation activity, were interpreted as evidence of the economic regulator's 'intellectual neglect' of the problem of long-term asset deterioration (House of Commons Environmental Audit Committee 2000, para. 208). The Competition Commission concurred with the Committee's analysis, criticizing 'Ofwat's heavy reliance on the serviceability criteria' and voicing its view is that 'more needs to be done to understand the relationship between asset condition and service ability' (*Utilities Journal* 2000*b*: 32). If correct, this argument lends weight to the charge that water companies, in anticipation of a rapid decrease in service levels due to declining infrastructure quality, may be seeking to dispose of assets.

In pursuing vertical de-integration, companies have put forward a variety of restructuring proposals (Ofwat 2000*c*). A number of companies considered the separation of ownership of the water-utility assets from operations and the introduction of a competitive outsourcing strategy for service provision (Ofwat 2000*d*). In its most radical form, the asset-owning firm would be entirely debt-financed, contracting out operations to independent service providers. In addition, some companies are considering whether a new structure of ownership might be appropriate for the asset-owning business, which would be owned not by shareholders, but by its customers (or a selection of them) or members in the form of a mutual or company limited by guarantee (or other not-for-profit vehicle). The two formal proposals put before Ofwat to date have entailed two distinct strategies: mutualization; and securitization.

The Kelda proposal: mutualization

The first formal proposal to Ofwat, by Kelda (the newly named parent company of Yorkshire Water) would have entailed mutualization of the core, regulated business.[19] In making its proposal, the company appealed to a long tradition of mutual ownership in Britain, and proposed to register the new

---

[18] Other agendas on the part of city analysts cannot be completely discounted; 3 of the (approximately) 30 City water analysts were actively seeking to become Conservative MPs in 1994–5 (in the run-up to the 1997 election).

[19] Another model that has been suggested is that of 'thin equity'—companies with a thin layer of participating shares, but not ordinary shares.

company with the Registrar of Friendly Societies.[20] Kelda depicted the proposed model—labelled RCAM (Registered Community Asset Model)—as a 'stakeholder' business, whereby assets would be returned to consumers, and consumers would benefit not only from increased control but also from the lower water prices which might result from the cheaper cost of capital available to the mutual.

The RCAM would have been a new company, owned by customers and operated on a not-for-profit basis, which would have acquired the assets and debts of Yorkshire Water and taken over its water supply licence. Wholly financed by debt on the bond market, the RCAM would have outsourced the management of the supply of water and sewerage services to a Kelda subsidiary for an initial period of 5 years, following which competitive bids from other water and sewerage companies for subsequent 5-year contracts would have been sought. This subsidiary, an operations management company, would have been unlicensed, and would have employed most of the former staff of Yorkshire Water. Under Kelda's proposal, any profits would be reinvested in the business or returned to consumers.

Some proponents of the RCAM model suggested that 'ownership of a regulated utility in this form would align the interests of the owners of the regulated businesses with those of its customers and in doing so would minimize political and regulatory risk' (Ofwat 2000*b*: para. 56). Others argued that the RCAM model represented 'a failed attempt to pass the assets and liabilities of an unsuccessful private business back to consumers' (Social Enterprise Institute 2001: 2). Given that Kelda was proposing the RCAM purchase Yorkshire Water's assets at what many industry observers judged to be an inflated price, and that the RCAM members—Yorkshire Water's current customers—would be responsible for the company's debt, the regulator commented that, 'although there appear to be clear benefits in the short-term for Kelda's shareholders from the proposals, in its current form, the benefits for customers have still be to demonstrated' (Ofwat 2000*d*: 4). The purchase price of the assets was one contentious issue: at £2.4 bn, the price was significantly higher than the 1989 sale price of £939.2m.[21] After subtracting debt of £1.4 bn., a net cash surplus of £1bn. would have remained with Kelda, which would have used the cash to finance new investment or buy back its own shares given that fact that it was no longer a capital-intensive group, thereby returning money to shareholders. Shareholders would thus receive something approaching the regulatory book value, while the mutual and thus customers could in the future take

---

[20] A mutual is a 'not-for-profit' corporate body set up for the benefit of members, who hold shares in the mutual. These 'shares' are different from those issued by a company limited by shares, as the maximum value permitted to be held by any one member is limited, and there is no requirement for the total number of shares to be limited (it is usual that no limit is set). Members are liable for any amounts unpaid on their shareholdings in the event of liquidation of the company. Profit is used in the manner prescribed by the rules of the mutual—typically, either reinvested or distributed among members. The Registrar for Friendly Societies registers mutuals and cooperatives under the Financial Services Authority.

[21] Director of Communications, Yorkshire Water, personal communication (Nov. 2001).

advantage of a lower cost of capital. Shareholders would also avoid future regulatory risk—particularly important given the new, tougher price cap regime in place since 1999.

Ofwat's response to the proposal was negative. The transfer of risk from shareholders to consumers posed several problems. While maintaining that 'the separation of asset ownership from operations and the outsourcing of operations could offer opportunities for greater efficiency resulting from the introduction of competition for the operations market' (Ofwat 2000*d*: 5), the regulator pointed to several concerns. First, would the new mutual have sufficient in-house, operational expertise to manage the business and supervise contracting? Would those responsible for the failures be clearly identifiable and have sufficient resources to bear the cost of their mistakes? Specific risks should be allocated to those best able to manage them. Equity markets were in this case to be preferred to customers; although the equity model results in a higher cost of capital, the risk to consumers was, in Ofwat's opinion, much reduced. Ofwat also pointed to the likelihood of Kelda continuing to win the operations contracts by virtue of superior knowledge of the network, and argued that 'the potential benefits to customers of greater outsourcing and a reduction in the cost of capital are not dependent on the introduction of a new ownership structure and could proceed under the existing equity model' (Ofwat 2000*d*: 5). The apparent inability of the proposed company structure to cope with unexpected cost shocks in the absence of an equity buffer, fears of poor quality corporate governance, and the lack of incentives to drive further efficiency gains in management, were other decisive factors in the regulator's decision to reject the restructuring proposal (Ofwat 2000*d*).

The Glas Cymru proposal: securitization

In rejecting the Kelda proposal, the regulator made clear that he was not opposed, in principle, to the idea of separating ownership and operation or new forms of ownership if this was seen to benefit the consumer, where 'benefit' was understood to mean an increase in competition and/or a lowering of prices without prejudice to efficiency incentives. The second proposal for a vertically de-integrated water supply business, put forward by Dŵr Cymru (formerly Welsh Water) some months after Kelda's proposal, was judged to meet these criteria, and was approved in January 2001. In contrast to the Kelda proposal, the Dŵr Cymru proposal entails securitization rather than mutualization,[22] creating a debt-financed, not-for-profit company owned by its members and limited by guarantee, rather than a customer-owned mutual. Glas Cymru, a new company limited by guarantee under the Companies Act 1985, owned by

---

[22] Securitization is the financing process whereby assets are sold to a new company, in order to be repackaged as marketable securities for sale to investors. In addition, Dŵr Cymru made significant changes to its business models and governance structure. Other companies have proposed securitization wihout these other changes.

its unpaid members rather than shareholders, was formed for the sole purpose of purchasing the assets from Welsh Water; the sale took place in May 2001. The Directors of Glas Cymru represent a variety of Welsh interests. The Directors have no financial interest in the company and do not receive dividends, but their financial remuneration is incentive-based; compensation includes a discretionary bonus related to service delivery performance compared to other water and service companies. At the time of writing, Glas Cymru had some 60 members (apart from the executive and non-executive directors) who were to be reappointed every 3 years, for up to 10 years.

Strong support of the Welsh Assembly was a key factor in the regulator's favourable decision, which was viewed as 'a key test of Welsh devolution' (*Utilities Journal* 2001*b*: 28). Equally compelling was the fact that Western Power Distribution, the new owner of Hyder Utilities (the parent company of Welsh Water), wanted to exit the water business and was willing to sell the assets at a discount on their regulatory asset value. Ofwat nonetheless imposed commitments (secured by changes to the company's operating licence) requiring good corporate governance, including: an incentive scheme for executive management; a prohibition on diversification; and the submission of procurement plans to Ofwat for review (Vass 2002). No other water company had been required to submit to these good governance requirements.

Another key factor in the regulator's favourable decision was Glas Cymru's promise to significantly lower consumers' bills.[23] Contracting out of operations to another company (United Utilities) would provide some savings. More important was Glas' financial programme; as the company noted, 'the water industry is very capital intensive and the cost of paying a return on money raised to finance assets is Welsh Water's single biggest cost, currently absorbing nearly a third of Welsh Water's annual revenues [i.e. customers bills]'.[24] The switch to 100% debt financing, through investment-grade bonds, entails not only a lower cost of capital, but also a greater surplus that can be invested in the network and in environmental protection, used to build financial reserves, or returned to customers. The advantages of Glas Cymru's lower risk profile, particularly given its commitment to remaining a *non*-diversified operating strictly as a regulated water business,[25] were confirmed by Standard & Poor's AAA rating of the company's main bond issue. Glas Cymru was the

---

[23] Although Ofwat did approve, in an interim determination, an increase in the price limits originally assigned at the 1999 Periodic Review, with a negative price limit only in 2000/1, and with prices rising above inflation every year after that (letter to MD of Dŵr Cymru/Welsh Water, 19 Dec. 2000, available on http://www.Ofwat.gov.uk/interims/final_interims/interim_wsh_final_19dec.html, accessed 29 Oct. 2001).

[24] Glas Cymru, 2001, 'Briefing on Membership', http://www.glascymru.com/english/pdfenglish/members/
BriefingMem.pdf Accessed 16 Oct. 2001.

[25] Glas Cymru is prohibited from diversifying into other activities, both by its constitution and by an undertaking to Ofwat that it will not change the constitution without first consulting with the regulator (*Utilies Journal* 2001*b*).

only water utility in England and Wales during the 2000–5 period to make voluntary price cuts (worth £23m.) on top of the agreed formula.

The 'people's company'?

What does the restructuring of ownership, management, and financing of water companies imply for consumer control of water company management practices and decision-making? In the case of Glas Cymru, the self-styled 'people's company', direct customer participation has not yet increased substantially (Taylor and Pickard 2001). The consultation undertaken by Glas prior to the restructuring was made up of eight focus groups, a survey of 'opinion formers', and a quantitative survey of 1,000 customers—out of 1.2 million households served by Welsh Water (Ofwat 2001c). In summarizing these customer consultation measures, Ofwat announced that it was 'satisfied that Glas has conducted as full a consultation as could be expected' (Ofwat 2001c).

Within the new company structure, direct customer participation mechanisms will not be substantively different. Accountability to Glas's customers will be ensured indirectly—via the Board being held 'accountable' to the Ofwat CSC and to the National Assembly for Wales. The importance of the support of the National Assembly for Wales was viewed as critical by Ofwat in its approval of the Glas proposal: in the absence of 'strong objections' on the part of customers, the support of the Assembly and the Ofwat CSC was a key factor in Ofwat's decision.

Indeed, direct customer participation was viewed with ambivalence by Ofwat, which noted a risk that 'members of Glas be captured by special interest groups' (Ofwat 2001c:9), and reiterated the need for 'arrangements for the appointment and functioning of the members of Glas designed to make sure that they remain focused on its commercial success' (Ofwat 2001c: 9). The key customer issues as outlined by Ofwat were not participation but rather: reduced efficiency, reductions in bills, and safeguards regarding the allocation of risk. Securitization, as undertaken by Glas Cymru, thus has not implied significant changes in direct water-user participation in company decision-making.

The ambivalence over customer participation in water supply management evident in Ofwat's decision on the Glas proposal was also evident in its decision to block the Kelda proposal for mutualization. Here, Ofwat's concerns centred on the risk of failing to attract the 'right calibre' of managers and to incentivize them, and a resulting weak management of competitive tendering, resulting in lower performance (Ofwat 2000c; 2000d; 2000e). Given the regulators' duty to encourage efficiency on the part of water companies, the possible reduction in efficiency incentives was also of concern. Specifically, the 'risk-averse' nature of a customer-owned mutual (where customers bear liability) would, Ofwat argued, result in the board being less likely to seek efficiency gains than would be the case with an equity-financed private company. The possibility that customers might legitimately have other preferences that might override the efficiency imperative did not appear to have been considered by the economic regulator. It is thus unsurprising that Ofwat has announced that the 'mutual' model is unsuitable for English water companies—as increased direct user participation is fundamentally at odds with the requirements of the economic regulatory framework—particularly the paramount duty placed on the regulator of incentivizing efficiency. Limited direct consumer participation in water company management is thus likely no matter what the form of the newly restructured water company: mutual, securitized private, securitized charity, or ring-fenced. This limited participation is attributable in part to the contradiction in the regulatory framework between the obligations placed on the economic regulator (i.e. efficiency gains and permitting companies to finance their functions) and perceived impacts of greater consumer participation (i.e. a decrease in efficiency incentives).

Restructuring financing

Following the approval of the Glas Cymru securitization, many companies expressed an interest in restructuring. AWG announced plans to restructure, and was rebuffed (although restructuring was eventually allowed to proceed). Given the explicit commitment on the part of the regulator for the existing equity model, backed by political support, Pennon has rejected restructuring for the moment. Severn Trent has also rejected restructuring, and plans to focus on the non-regulated side of the business, reinventing itself as an environmental services company; growing to a 10% share of the UK waste-management business by 2001 (Global Water Intelligence 2001). United Utilities plans to diversify further into the non-regulated parts of its multi-utility business.

Following the success of Glas's strategy, other firms are adopting debt-oriented refinancing strategies. In March 2001, one of the small independent water holding companies, East Surrey Holdings (owners of the water-only company Sutton and East Surrey) announced plans to change its structure in search of reducing the cost of capital. They have set up a credit-wrapped index-linked bond. This is a different form of debt finance, which enables the borrower to repay the debt at a very low interest rate. The difference between the real interest rate and the agreed lower interest rate is added on to the debt. Sutton and East Surrey hope to reduce the cost of capital from the 7.3% anticipated in their 2000 plans to around 4.5%. Three of the largest of the water and sewerage companies (Anglian Water, Southern Water, and Wessex Water) are planning to tightly 'ring-fence' asset ownership from operations within the group-company structure and refinance the ring-fenced utility via debt rather than equity. Debt-financing is also being employed by Mid-Kent, another smaller, independent water-only company which undertook a management buyout in 2001, combined with a ring-fencing of its core operations (Ofwat 2001*b*).

Water companies have justified plans to restructure financing by arguing that equity was an expensive source of finance, and that other sources of finance—in particular debt finance—were more viable in the long-term. Supporters of privatization have argued that sourcing investment from equity, although more expensive than government debt, creates pressure on managers to make efficiency gains which offset the increased cost of capital. Opponents of privatization argue that debt, and in particular government debt, is so much cheaper than equity that any efficiency gains under private ownership would not be outweighed by an increase in the cost of capital. Twelve years after privatization of the water industry, many water companies appear to have taken the latter view, bringing into question one of the justifications for privatization—that equity markets, because of the scrutiny and pressure to which they would subject managers, thereby driving efficiency gains—were preferred sources of finance, for the water industry.

Is restructuring the solution?

Restructuring has entailed innovation with ownership, company structure, and financing. Three different ownership models have emerged: private for-profit; management-owned for-profit; and member-owned non-profit. Five different organizational models have emerged: a diversified, but integrated 'environmental services' business with regulated water and non-regulated waste and similar operations (for example Severn Trent); a UK-based regulated business supporting an international water business (for example Thames); a UK-based multi-utility business (for example United Utilities); a ring-fenced, highly geared UK water business within a wider group (for example AWG/Anglian); a non-diversified UK-based company (for example Glas Cymru) (Rathbone 2002). In addition, some of these organizational models entail contracting out, with asset-owning firms sub-contracting service management—similar to the French model of concession and leasing contracts for water supply infrastructure management. Bond markets have supplanted and in some cases replaced equity markets as main sources of financing; retained profits, which could provide another important source of finance, have been constrained by the 'dividend machismo' of the industry which according to some City analysts raised the cost of equity, lowered investment, and destroyed long-term value (ironic, given that this 'dividend machismo' was driven, at least in part, by the demands of City analysts).[26]

[26] R. Miller-Bakewell, personal communication (May 2002).

To date, no companies have followed the example of Glas Cymru's 'securitized charity' model. The hesitancy of companies to replicate Glas's model has arisen, in part, because Ofwat has warned that it 'does not see the [Glas Cymru model] as a model for the industry as a whole' (Fletcher 2001a). Given his duty to encourage competition, and the great technical difficulties posed by common carriage, the regulator might be expected support competition in the production chain (which would entail vertical de-integration of water companies, and entail another round of consolidation in the industry). However, this conflicts with concerns about the ability of companies financed by debt rather than equity to make improvements in efficiency in the absence of shareholder pressure. As noted by one of the principal architects of privatization, shareholder scrutiny is highly effective because it provides accurate, rapid assessments of company performance through the movement of share prices, as shareholders make direct comparisons between the performance of different companies, which 'ensures that each [company] is under pressure to run its entire business efficiently and innovatively' (Littlechild 1986: 2). The difficulty with the regulatory framework, from the political perspective, is that efficiency gains do not necessarily translate into lower prices, or into investment in long-term maintenance of water supply infrastructure; shareholder pressure has acted against discretionary rebates and against discretionary CAPEX. The implications, particularly with respect to the integrity of Britain's water supply infrastructure, will probably not be known with any certainty for decades.

Political support for the equity model, given the government's desire to avoid the failure of privatized utility companies, particularly in the context of the bankruptcy of Railtrack in the autumn of 2000, remains strong. Given that pressure for water industry restructuring is not coming from consumers, but from 'factors lying largely within the expert communities responsible for or concerned with the industry' (Summerton 2001: 22), the regulator is not likely to approve other restructuring proposals which entail divestiture of assets in the near future. This serves merely to delay the drive to restructure, rather than resolve the contradictions of the post-privatization regulatory framework besetting England's regulated private water companies.

The contradiction between the short-term horizon of financial markets and long-term nature of water industry investments has been slightly alleviated, but not eliminated, by the shift to bond financing. The inability of short-term periodic review periods to facilitate long-term planning in the industry remains a concern. Regulatory risk has been reduced by the greater transparency of the new economic regulator[27] (appointed after the 1999 Periodic Review), but fears about government intervention remain. Efficiency incentives may be reduced in the new debt-financed and/or ring-fenced water companies.

---

[27] The first regulator, Sir Ian Byatt, was succeeded by Philip Fletcher upon Byatt's retirement in July 2000.

Ring-fenced companies may also lose the synergies between the regulated and unregulated businesses within the diversified groups to which most water companies now belong; without incentives to adopt strategies benefiting both the regulated water business and the group, tensions may arise (Rathbone 2002). Debt-finance may not lower the cost of capital over the long-term. Two years after the review industry analysts were continuing to predict lower dividend growth, and a drop in earnings per share on the part of all but one of the water companies (Miller-Bakewell 2001*a*). This arose in part because the forward investment programme is largely non-negotiable, stemming in most cases from legislative requirements to meet specific water quality or environmental standards. The chief challenge going forward will be complying with the requirements of the new European Water Framework Directive by 2010–15.

Conclusions

Those who assert the success of the regulatory framework point to the achievement of requiring companies to lower prices while delivering quality improvements; from this perspective, an effective balance has been struck for the price–quality trade-off. The regulator has argued that price cap reductions allowed the benefits of efficiency gains to be returned to customers. Following the Periodic Review in 1999, however, water company managers have argued that the regulator has prioritized his duty to customers in the short-term, to the detriment of his duty to ensure that the industry is able to finance its functions in the long-term. In the absence of a balance between these two duties, the first decade following privatization was characterized by profits that the regulator characterized as 'excessive', but the current decade will be characterized by increasingly heavy capital expenditure requirements in a much less forgiving financial environment. As explored above, continued restructuring is the likely outcome.

Vertical de-integration and mutualization are the most significant changes in company structure since privatization in 1989. Restructuring calls into question whether the model created under the original regulatory framework—vertically integrated and equity-financed private monopoly suppliers of water services—is delivering the lowest prices for consumers. Companies argue that debt finance is cheaper than equity finance, and that efficiency incentives created by the presence of shareholders (via equity financing) do not outweigh these savings. Furthermore, the lowering of risk in a non-profit, non-diversified structure in turn lowers the cost of capital—with the potential of significantly lowering prices. Companies themselves are thus challenging a central premise of the British model: that the regulatory framework would harness the profit motive of privately owned equity-financed companies to drive efficiency gains and thus lower prices.

What is implied by this departure from the vertically integrated equity-intensive private monopoly model created in 1989? Supporters of privatization have argued that sourcing investment from equity, although more expensive than government debt, creates pressure on managers to make efficiency gains which offset the increased cost of capital. Opponents of privatization argue that debt, and in particular government debt, is so much cheaper than equity that any efficiency gains under private ownership would not be outweighed by an increase in the cost of capital. Twelve years after privatization of the water industry in 1989, some water companies appear to have taken the latter view, bringing into question one of the key justifications for water privatization.

In the longer-term, the majority of the remaining independently listed companies are likely to continue to diversify and/or internationalize into non-core activities, or employ another model of vertical de-integration. Alternatively, some parent companies may choose to load debt onto the core water supply company, weakening the balance sheet, and allocating high risks to the regulated business. Others may choose to sell the regulated water business—if buyers can be found. At the time of writing, the French company Lyonnaise des Eaux/ONDEO (one of the two largest water companies in the world) had recently announced its intention to sell Northumbrian Water, its flagship subsidiary in England, citing low growth potential, a reduced rate of return, and the strict regulatory environment to which Northumbrian's activities are now subject (Orange 2003).

In pursuing restructuring strategies, water companies are responding to the dilemma faced by all of the remaining asset-owning water supply companies: generating sufficient revenues from the regulated business, in a capital-intensive industry with stagnant demand and limited market growth potential, is predicated upon a growth in prices which, for a partially non-substitutable resource essential for life, raises questions of equity and access and results in public and political pressure to keep (monopoly) prices and rates of return down. Restructuring strategies to date have evaded rather than resolved this dilemma.

8

At the Frontier of the Market? Re-regulating Water Supply

> If you can privatise the water industry, you can privatise anything. The government reckons it can (at least in England and Wales).
>
> *The Economist*, 8 Feb. 1986, p. 27

The decision to privatize the water supply industry was a seminal moment in the British government's privatization programme. Earlier privatizations had concentrated on those industries which were clearly profitable, some of which even had private competitors. To a greater degree than any of the industries privatized before it, water lay at the frontier of the market—so much so that new regulatory institutions had to be designed and regulatory bodies created in order to enable the market to function.

Previous chapters have explored the contradictions and trade-offs embedded in this regulatory framework. Successive economic regulators have struggled to maintain a balance between the need to generate stable, sufficiently high levels of return to satisfy shareholders on the one hand, and politically acceptable rates of return, on the other. Charged with the duty of increasing competition in the water industry, regulators are faced with a trade-off between maintaining a sufficient number of comparators for comparative competition to function, and maintaining sufficient 'takeover' pressure on water company managers as an incentive for performance. With respect to water pricing, regulators are confronted with a contradiction between the outcome of the application of the principle of economic equity (spatial differentiation of prices to facilitate competition), and a politically acceptable threshold of spatial homogeneity of bill levels. In environmental regulation, a contradiction may arise when employing techniques of water valuation: an accurately valued environment may not be sufficiently valuable enough (in monetary terms) to justify the goals and standards of environmental conservation. When implementing demand-side management policies in the water industry, a (narrowly defined) concept of economic efficiency as employed in long-run marginal cost pricing formulas may in some instances counteract water conservation, in the sense of a sustained reduction of water demand.

As detailed in earlier chapters, these contradictions have to some degree undermined the British model of vertically integrated, equity-financed private water companies, regulated through price cap controls and comparative competition. The failure to balance these contradictions—and the ensuing volatile mix of high profits, large dividends, rising leakage rates, rising prices, and environmental raiding during droughts—provided the political justification for re-regulation of the water industry. The scope and scale of environmental and economic regulation have constrained water companies' activities to a degree unanticipated by the architects of privatization. Re-regulation has, in turn, contributed to the ongoing restructuring of the water supply industry. What are the implications? What lessons can be drawn from the British case, for countries seeking greater investment in their water services, considering the involvement of the private sector, yet also concerned with equity in service provision? This chapter comments briefly on these and other questions of relevance to those grappling with the issue of private sector participation in water supply.

The demise of price cap regulation?

The focus of the architects of the regulatory framework was on efficiency incentives, and on levels of service and prices for given water quality standards; from the companies' perspective, investor value and customer service were the two primary interests which managers were required to balance. Additional objectives—such as technical efficiency (e.g. leakage rates), social equity (e.g. affordability), and supply system security (e.g. headroom)—were originally downplayed in the regulatory framework. The key characteristics of the original design of RPI − X which made it preferable to American-style rate of return regulation were its simplicity, the lower burden imposed on regulators and government, and the associated lower cost (Littlechild 1986). Littlechild recognized the existence of an effectiveness–comprehensiveness trade-off, whereby an increasing the number of regulations will lend less weight to any one measure, making the regime more complicated to impose, increasing the cost of enforcement and decreasing its effectiveness. Regulation with a 'light touch' was necessary because 'effective regulation requires simplicity' (Littlechild 1986: 6).

In its original form, then, price cap regulation was intended to entail a light regulatory burden, with relatively small information requirements; companies' price limits were to be set by the regulator once every ten years, and the opportunity to retain profits within the price limits, backed up by pressure from shareholders and the City, would provide the incentive for efficiency gains. Yet only a few years into privatization, key variables upon which the forecasts underlying the price limits had originally been based had changed substantially; the regulator decided that an 'interim' price review should be held, in order to readjust price limits. The decision to carry out Periodic Reviews at 5-year intervals hints at the difficulty of accurate forecasting. The resulting shortening of the review period, combined with instances of regulatory pressure in intervening years on companies not to take up full price caps, has meant that the original incentive structure of price cap regulation has been substantially degraded, sparking pronouncements of the 'death' of price cap regulation by regulatory economists:

> a conflict between stable returns at the industry level and individual incentives lies at the heart of the problem with the current regulatory system. The . . . political and commercial unacceptability of abnormal industry-wide profits [or] losses means the current system of price cap regulation will almost inevitably break down. Whether or not price cap regulation is formally abandoned in the near future, interventions by the regulator between reviews will undermine the operation of regulation as anything like pure price cap. (Mayer 2001: 10)

The impact of the price cap reduction, and earlier interventions by the regulator to prevent companies from taking up their full price cap, calls into question the robustness of the efficiency incentives as originally designed (and perhaps indicates that the architects of the original regulatory framework underestimated the difficulty of assessing costs—an essential input into the price cap calculation). The regulator notes that 'the existing . . . model has led to greater efficiency of the water companies . . . [and] the 1999 price review transferred the benefit of this increased efficiency to customers' (Ofwat 2000*d*: 4). But the fear of 'claw-back' of profits that this instils in companies undermines the incentive to maximize efficiency. The political unacceptability of higher profits and commercial unacceptability of high losses introduce a contradiction between the safeguarding of the efficiency incentive, and the preservation of stable and politically acceptable rates of return.

Water companies are now more tightly regulated than perhaps any other of the privatized industries in Britain. To some extent, this is linked to a broader change in political direction in Britain. The transformation of the Conservative's 'Deregulation Unit' to a 'Better Regulation Unit' within the Cabinet Office after Labour's election victory in 1997 is more than merely semantic. The water industry, however, remains a special case. Although other utility network industries regulated by price cap (gas, electricity, telecommunications) had experienced some degree of 'regulatory creep' post-privatization, significant portions of these industries had been opened to competition, with a consequent reduction of regulatory activity—even the phasing out of price caps in some cases. In the case of water, the introduction of competition has been very limited, and the government has scaled back on original plans for competition in the domestic sector. Regulation, as such, has not 'withered away' in the case of the water industry; quite the contrary. Yet, as detailed in previous chapters, price cap regulation is now something very different than the architects of the post-privatization regulatory framework had envisaged;

most importantly, it now entails explicit decisions about the rate of return—announced by the economic regulator in the last Periodic Review to be just under 7% (pre-tax) for Britain's water utilities. To the degree that the principle of capping prices, rather than rates of return, was central to the original economic regulatory framework, pure price cap regulation has ceased to be applied in the British water industry.

The retreat of the market?

The current wave of restructuring in the British water supply industry should not simplistically be read as a retreat of the market. First, these proposals leave unchallenged the distributive theory of justice implicit in the privatized, commercialized model of service provision implemented in Britain, which prioritizes access to material goods as a means of need-fulfilment. The process of commercialization, and the impacts of this 'delocalization' of companies on resources management and responsiveness to users, are unaffected by the restructuring proposals (Page 2002). On the contrary, the commercialization of the core business has been further entrenched by the progressive diversification of the water companies into non-core, non-regulated activities. Nor do the restructuring proposals substantively roll back the commercialization of relations between water users. Commercialization implied the rescripting of consumers as customers rather than citizens, a deliberate de-politicization of water regulation through the creation of arm's-length regulators, and a shift from social equity to economic equity in water pricing and hence consideration of willingness but not ability to pay—with the burden of increasing water bills falling disproportionately (in terms of income and access) on the most vulnerable consumers (Bakker 2001). The substantive participation of consumers in water policy-making has not been significantly altered by water industry restructuring (Page 2002). The resistance of consumers and politicians to commercialization, which was articulated with a change in policy (and also government) at the national level (Leys 2001), drove the re-regulation of the industry, which was an enabling condition for, rather than result of restructuring.

Nor do the restructuring proposals substantively alter the transformation of users' entitlements entrenched by the post-privatization regulatory framework. The contested politics of redistribution post-privatization hinge upon a tentative, and contested double discursive move: reconfiguring citizens as consumers, and the environment as a legitimate user whose interests are to be balanced with those of consumers. Post-privatization, decision-making on capital investment in the industry balances the interests of consumers 'willingness to pay' against environmental protection and rehabilitation requirements—a cost-benefit exercise which minimizes the participation of labour and attempted to exclude questions of ability-to-pay, in distinct contrast to pre-privatization (Bakker 2001). The majority of increases in domestic water users' bills post-privatization have gone towards the cost of improving not only drinking water quality but also beaches, bathing waters, and inland waters through investment in waste water treatment facilities, mandated by EU directives and backed up by local pressure from campaigning NGOs such as Surfers Against Sewage (Ward 1996). Over the past decade, river water quality and drinking water quality in Britain have improved significantly; river water quality appears to be at its highest level since the Industrial Revolution (DEFRA 2001*a*; DWI 2000; EA 2001). However, the implications of continuing environmental improvement for water prices are highly contested. The reduction in prices at the 1999 Periodic Review implied a sharp reduction in the size of the investment programme, as Ofwat rejected most of the companies' plans for increased 'environmental spend', citing the need to reduce the cost burden on consumers. Distributive struggles over water continue to be articulated through a governance model which prioritizes market-based adjudication of the interests of consumers, investors, and the environment, in contrast to the public policy-led mediation of the interests of citizens, business, and labour which characterized the nationalized industries. The core of the new water supply governance model introduced by the commercialization and privatization of the water supply industry remains intact.

Water politics

Some politicians hope, naively perhaps, that devolving operational responsibility for water supply will allow them to devolve political responsibility for water supply delivery; this is often an important reason for the attractiveness of private sector participation in water supply to political leaders. 'Depoliticizing' water supply delivery is also an oft-stated goal of water supply managers seeking greater autonomy from public oversight (in some cases, for very good reasons). Economists also argue that social policy, and political imperatives (such as employment creation) should be removed from water policy. The result, it is argued, will be a politically effective system—equitable, transparent. The original architect of the price cap framework, for example, argued that RPI − X would be 'politically effective' because it 'focuses attention on what consumers ultimately care most about: the prices they will have to pay' (Littlechild 1986: 27). Yet, as explored in Chapters 5 and 6, water pricing policy post-privatization has been politically contentious; a central concern has been the perceived lack of mechanisms to address the regressive distributional effects of water privatization on customers. Additional concerns arose about safeguarding operational efficiency, supply system security, and environmental protection when this conflicted with the goal of economic efficiency.

Inevitably (and not in all cases unwillingly), politicians were drawn into this debate, and played a significant role in the re-regulation of the water supply industry.

That consumer concerns were so central to the post-privatization water management debate may seem surprising if one considers the relatively limited role granted to direct consumer participation in water policy-making (Page 2002). The functional separation of responsibility for economic criteria, environmental standards, and drinking water standards between three separate regulators operating at 'arm's length' from both government and water companies has resulted in a complex regulatory framework. Centralization and the complexities of the issues have further facilitated the creation of a 'management structure that is inaccessible and exclusive' (Strang 2001: 61). Yet water policy-making is nonetheless increasingly pluralistic and participatory. To paraphrase two political scientists, regulation has been transformed from 'the private management of public business to the public management of private business' (Richardson, Maloney, and Rudig 1992; Maloney and Richardson 1995). The number of stakeholders effectively intervening in water policy decision-making, and the extent to which they intervene in policy decisions has increased. Within this context, the influence of certain stakeholders (e.g. organized labour) has substantially diminished whereas other groups have increased their influence—particularly special interest groups at the national level, notably environmental NGOs and also consumer lobby groups (Maloney and Richardson 1995; Page 2002).

Affordability of bills was one concern voiced by these groups—and by other key participants in the water supply debate, including (then-opposition) Labour politicians. Excessive dividends, 'fat cat' salaries, and 'water poverty' resulting from rapidly rising water bills were primary concerns on the part of the media and the public (see, for example, Graham and Marvin 1994; Helm and Rajah 1994; House of Commons Environment Committee 1996; Graham 1997; Guy, Graham *et al.* 1997; Marvin and Guy 1997; Ward 1996). Resource management also received increasing attention, in part because privatization coincided with the beginning of a period of extreme meteorological and hydrological variability, and in part because of the negative publicity stemming from the management of water supply during the Yorkshire drought. Key issues included leakage rates, forecasts of large structural supply deficits, and new resource developments and their alternatives. Associated debates focused on the structure of regulation, the feasibility of competition, company management practices, and the integration of UK environmental policy-making with that of the European Union.

The concerns voiced in this debate have been an important justification for the substantial re-regulation of the water industry which has ensued: the reduction of price caps; and the re-imposition of 'command-and-control' type of legislation and regulations. In practice, water company managers are highly constrained. Limits are imposed on management freedom to: set prices; decide about investments; decide which raw materials to develop; decide which services to provide; vary service quality; refuse supply; merge, divest, and restructure; and to dispose of assets (Rees 1998). To some degree it is unsurprising that this should be the case. The government has divested ownership and thereby financial responsibility for the industry, but has maintained control over standard setting, which it regulates stringently given the relatively high degree of natural monopoly and public good dimension of externalities associated with water supply. Re-regulation should thus be understood as a process of political negotiation, as well as organizational change, which implies that privatization and commercialization should not be read (simplistically) as an increase in the dominance of private interests, any more than nationalization should be read as a diminution of those interests.

Regulators and government have thus been drawn into increasingly interventionist roles in order to require companies to achieve desired social goals (such as water conservation and protection of vulnerable consumers). Despite the original vision of the architects of the regulatory framework who hoped that water policy could be denuded of many social policy goals, questions of social equity and access continue to figure centrally in water policy debates. Any regulatory framework for privatized water supply should include mechanisms that adjudicate between economic and social equity, and between (narrowly defined) economic efficiency and environmental protection. The failure to balance these goals was a causal factor in re-regulation, and hence restructuring of the industry.

The regulatory framework ran up against a central political economic contradiction of a privatized water supply industry: maintaining a sufficient rate of return in a highly capital-intensive industry with extensive public health and environmental externalities requires price levels that will be politically contested, and in some cases politically unacceptable. This contradiction will resurface, as water supply is a partially non-substitutable resource essential for life—a fact which regulators and governments ignore at their peril. Questions of equity and access can only be displaced, not dispensed with, by commercialization and privatization; water supply remains an inescapably political subject.

Exporting the British model?

At the time of privatization, advocates had high hopes for the success of the model domestically, and made confident predictions about the export of this model abroad. Yet more than a decade after privatization, the English and Welsh privatized water model of large vertically integrated private monopolies regulated by yardstick competition remains unique. Few governments have divested itself of water infrastructure assets, and none of the resulting private companies operate on the geographic, demographic, and financial scale of the largest English and Welsh water and sewerage companies. A few countries and municipalities have divested itself of water supply companies (for example Santiago de Chile); the majority has opted for concession contracts to private companies, operating assets owned by, or eventually to be turned over to governments. Even the 'World Bank is believed to favour the French over the British variant, on the grounds that it retains the public ownership option, and keeps a sanction against an unsatisfactory operator' (Winpenny 1997: 4). The sophisticated regulatory model requiring the collection and analysis of large amounts of data is also a discouragement to governments seeking to privatize water services, although forms of 'comparative competition' have been adopted in some countries with multiple concession contracts (for example Chile). Other arguments against the English regulatory framework note the political unacceptability of an abrupt transition from public to private asset ownership, and the existence of relatively few existing large-scale water supply networks whose divestiture would attract willing buyers. In the global marketplace for governance models, products and services, the British model of water privatization is at a disadvantage. It may also be at a logistical disadvantage, given that English companies operating overseas are finding it harder to source trained expertise, as the cohort of domestically trained managers shrinks due to retirement and downsizing, in contrast to their much larger French competitors, particularly Lyonnaise des Eaux/ONDEO, and Vivendi Environnement.

The exception may be the relatively undeveloped water services markets in some other OECD countries, particularly North America, where the water services market is large (an estimated US$55bn. USD per year for the USA alone, in contrast to the English and Welsh market, with turnover at less than US$10bn. in 1998/9),[1] relatively fragmented (run largely at the municipal level) and with the vast majority still under public ownership (Miller-Bakewell 2001*b*). In the United States, investor-owned companies are already subject to rate-of-return regulation, whether under domestic or foreign ownership. In both the United States and in Canada, outsourcing rather than divestiture is the most likely source of market expansion; the 'French' rather than 'English' model is much more likely to be pursued, given the fragmentation of the market. Given the fact that economic regulation of privately owned utilities is localized in both the United States and Canada at the state or provincial level, water regulators in North America may eventually take up some of the 'comparative competition' features of the UK price cap model; indeed, many energy and telecommunications sector regulators around the world have already done so.

Given the controversial expansion of privatization and private sector participation in the water supply sector overseas, the evolution of the English 'model' of regulation (increasingly applied to concession contracts in the

---

[1] WSA 2000, table 6.2. Exchange rate assumed: US$1.45 = £1 sterling.

South) merits close attention. In particular, the original assumptions of the British model of regulation—that the full costs of water provision can be devolved onto customers, that prices can rise to fully cover environmental externalities, that water is sufficiently similar to other network utilities such that a standardized price cap model of regulation can be applied, and that light-touch regulation is both feasible and politically acceptable—deserve careful scrutiny.

A mutual future?

The notion of decentralized, participatory community-led water management models is receiving increasing attention in developing countries, for irrigation as well as for water supply, as a means of enacting decentralization, and as a response to the perceived failings of the 'state hydraulic' paradigm (Gleick 2000). Within this more generalized and ongoing discussion have emerged new trends in governance models. One such model is 'associative self-governance'—a revived, rather than novel model for socio-economic organization, particularly of community services. The proponents of associative self-governance subscribe, if implicitly, to a 'community' model of governance, in which collective as well as individual incentives may be created under alternative ownership and management structures which resolve the trade-offs between shareholder and customer interests evident in the case of privatized monopoly services (Birchall 2001; 2002; Holtham 1997; Kay 1996*b*; Morse 2000). This position contrasts with the market model of governance which underpinned the regulatory frameworks created at the time of utility privatization. Proponents of the market model of governance assert that neither governments nor consumers should be involved in operational or management functions, and categorize consumers as customers rather than citizens (as under nationalized or state-led governance models), or empowered users.

In Britain, some advocates of associative self-governance support the creation of community 'mutuals' to run public services (see, for example, Mayo and Moore 2001 and the debate in www.themutualstate.org). The idea of giving consumers of services such as health, education, and utilities more control over service provision has received significant and growing attention within the UK, including high-profile support from some Labour MPs and government ministers. In a political context in which mutualization is actively being explored by senior government ministers as a means of concretely implementing Labour's 'Third Way', the Labour Party's recently released National Policy Forum consultation document on health and social care, for example, makes explicit reference to the creation of 'mutuals' or 'public interest companies' within the NHS (Labour Party 2002). Other alternative ownership and management structures are being actively pursued. Following the collapse of the privatized railway infrastructure provider, Railtrack, the government has decided to implement a not-for-profit public trust in its place. Similarly, in the London borough of Hackney, the failure of the local education authority led not to a private for-profit alternative, but instead to the creation of an independent not-for-profit trust.

Within England and Wales, water supply provides a concrete case through which to interrogate the attraction of the 'mutual alternative'. 'Mutuals' take many different forms, and not all imply greater participation of ordinary citizens, nor are they incompatible with continuing commercialization of public services, whose impact should be examined. Proposals to return water supply infrastructure to public control through the 'mutual' model have attracted a great deal of interest given the innovative and influential British model of water supply privatization, and the rapid growth of privatization and private sector participation in water supply around the world in the past decade. Some analyses have depicted the restructuring proposals as a 'retreat' from privatization, and as a reassertion of the 'commons' or the community over the 'commodity' property relation.

In contrast, I would argue that it is important to examine the degree to which alternative ownership and management structures under consideration entail changes in the 'commercial' governance model implemented in 1989. This is an analysis of the restructuring of water supply companies in England and Wales within the generalized trend of the erosion of traditional bases of political power in the advanced industrialized democracies over the past couple of decades, and associated restructuring of states. Within the so-called shift from 'government to governance', formal state authority is supplemented or supplanted by increasing reliance on informal authority, particularly in forms of negotiated patterns of public-private-community cooperation. Roles previously allocated to governments are now increasingly and controversially categorized as more generic social activities which can be carried out by political institutions, but may also be carried out by other actors (Pierre 2000).

Here we can situate the restructuring of the water industry within broader debates over 'associative self-governance' taking place in Britain. Chapter 7 argued that the failure of the post-privatization regulatory model to contain the contradictions between stable returns and the efficiency imperative, on the one hand, and politically acceptable rates of return and the equity imperative, on the other, led to a re-regulation of the water supply industry. This process of re-regulation was a key factor in restructuring, which has entailed multiple strategies (diversification, internationalization, vertical de-integration, mutualization, securitization), which were briefly analysed in turn. In contrast to analyses which depict restructuring as a 'retreat of the market', the chapter emphasizes the continuity of the commercial governance model applied in the water supply industry since 1989. In interpreting restructuring as an industry response to re-regulation of services provision, it is important to interrogate the incentives underpinning current moves towards a 'mutual' future.

In particular, proposals to mutualize arising from within the water supply industry should not be simplistically read as a retreat from the market. First, these proposals leave unchallenged the distributive theory of justice implicit in the privatized, commercialized model of service provision implemented in Britain, which prioritizes access to material goods as a means of need fulfilment. The commercialization of the industry began in the early 1980s, when the Water Act (1983) and supporting legislation initiated and formalized the transformation of the water industry in Britain 'from a public service to a business organization' (Penning-Rowsell and Parker 1983: 170). By the late 1980s, the nationalized industries were best characterized as 'publicly regulated private monopolies operating on modified market principles' (Hay 1996, 53). Privatization and the implementation of a 'price cap' regulatory framework consolidated this transformation. This commercialized service stands in distinct contrast to that of the nationalized era, during which access to a water service was regarded as a precondition of participation in collective social activity; an entitlement, extended to all.

None of the restructuring proposals to date challenge the progressive commercialization of the industry, initiated with the 1983 Water Act and entrenched by privatization, positioning water as a business rather than a service, with the goal of profit maximization rather than service provision (Bakker 1999a). Water remains classed in a category with other utility network commodities, for which the efficiencies generated by private ownership and exposure to the discipline of competition must be balanced by a restricted sphere of regulation by quasi-autonomous government agencies. Unlike more radical calls made internationally for a consumer-labour-environment alliance in restructuring water supply management (such as the Water Supply and Sanitation Collaborative Council's 'Vision 21', the Cochabamba Declaration, the Group of Lisbon's Water Manifesto, and the Declaration of the P7 at their 4th Summit in 2000),[2] current proposals for mutualization and customer corporations for the water industry in England would not significantly disrupt the governance model underlying privatization. This does not imply that the restructuring proposals pursued to date in the water supply industry will be unsuccessful, but contests interpretations of these models as a radical reworking of public services governance in Britain.

[2] The Water Supply and Sanitation Collaborative Council, located in Geneva, is a non-profit organization with funding from governments and multilateral agencies that acts as an 'international policy thinktank' on water management. The Cochabamba declaration followed a meeting of several hundred people in this Bolivian city concerned about the involvement of private sector corporations in water supply management. See http://www.canadians.org/blueplanet/cochabamba-e.html. The Group of Lisbon is a group of academics and policy makers from around the world which analyses globalization, calls for new economic governance models, and promotes new forms of social contracts. See Riccardo Petrella (2001), *The Water Manifesto* (London: Zed Books). The P7 (now P8) annual conference was convened for the first time in June 1997 by the Green Group in the European Parliament, as an alternative Summit to the G7 (now G8). Representatives from the world's poorest countries attend the conferences, which focus on the structural causes of and solutions to poverty. The declaration of the 2000 meeting countered the Inter-Ministerial Hague declaration earlier that year, and spelled out four principles for water democracy.

Water supply management: equity versus justice?

Britain must leave behind the sterile, century-long conflict between enterprise and fairness—between the left, which promoted the good society at the expense of the good economy, and the right, which promoted the good economy at the expense of the good society, and too often achieved neither. (Gordon Brown, UK Chancellor, 10 Nov. 1999[3])

The evolution of water charging policies in England and Wales over the past three decades has been underpinned by a shift in the prioritization of equity from social towards economic equity, and from the ability to pay principle towards the benefit principle. As discussed in Chapter 6, a goal of economic equity (justified on the basis of increased efficiency) implies the spatial de-averaging of water charges and cost-reflective tariffs. In contrast, a goal of social equity (justified on the basis of public health implications) implies cross-subsidies and wealth-reflective tariffs. Given the incomplete application of policies of equalization, and the incomplete penetration of metering and application of marginal cost pricing, neither principle of equity has been fully applied in practice. The current consensus—that universal metering is theoretically desirable but impractical and expensive, and that spatially de-averaged charges and fully cost-reflective tariffs are politically unacceptable—implies that temporal and spatial cross-subsidies will continue in the water sector. The emphasis on economic equity has shifted the balance of these cross-subsidies. Increasingly stringent regulation, and the direct intervention of the government on numerous occasions in the late 1990s have sought to mitigate some of the most politically unacceptable effects. The pursuit of economic equity, at the expense of social equity in water charging policy has thus been somewhat tempered.

The drive to commercialize and the underlying attraction of the goal of economic efficiency must be understood in the light of declining labour productivity in Britain over the second half of the twentieth century; calls for improvements in productive efficiency across all sectors have been a mantra across the political spectrum. The degree to which the 'state failure' hypothesis has taken hold is perhaps best exemplified by Labour's interventions in the water industry—strikingly limited when compared with its calls to renationalize the industry as late as 1992. Labour's policy interventions in the water sector since 1997 have been characterized by: a continued commitment to privatization, increased competition, and greater consumer choice; higher but less steeply increasing prices mitigated by alternative charging schemes for vulnerable groups; and more stringent regulation of companies, particularly with respect to profits, performance, and water conservation.

The motivations for this approach differ less significantly than one might assume from those of the architects of privatization. Water remains classed in

---

[3] Pre-budget statement, 9 Nov. 1999, Hansard, cols. 884–91.

a category with other utility network commodities, for which the efficiencies generated by private ownership and exposure to the discipline of competition must be balanced by a restricted sphere of regulation by quasi-autonomous government agencies. The abandonment of the categorization of water as a strategic resource by the Conservatives appears to hold true for New Labour. Consistent with the ideology underlying the privatization of the utility industries, individuals are treated as customers buying a commodity, rather than as citizens entitled to a service, although Labour has introduced greater controls on company behaviour, higher service standards, and greater protection for vulnerable groups and low-income consumers.

The Labour government's innovation within this paradigm is perhaps best characterized as a task of moral resuscitation, and a mitigation of the least politically popular outcomes of privatization, not an attempt to reintegrate water within a separate sphere of justice administered by the welfare state. Labour's justification for these alterations to the form, if not the substance, of water regulation stems from a moral duty:

Water is essential for life and health. Access to a sufficient supply of clean drinking water is fundamental to the well-being of citizens. The water charging system can and should make a major contribution to tackling social exclusion, supporting those in greatest need. (DETR 1999*b*: 11)

This terminology of basic needs, rather than of rights or entitlements, displays significant continuities with the approach to water poverty in the years following privatization. Individual water companies responded to the plight of low-income families or those with special needs by setting up water charities, to which those requiring exemptions or special treatment were required to prove their eligibility. Labour's water charging proposals formally integrate this charity function into charging mechanisms, rather than in voluntary company-administered rebates or discounts for consumers. However, the government has not sought to redress or restrain interregional price differentiation. Nor have the more regressive elements of the water charging systems, such as the payment of environmental improvements through water charges (which impacts most highly on low-income households, as opposed to more fiscally progressive taxation), been addressed. This is perhaps consistent with New Labour's broader approach to public services and evolving approach to regulation (Monbiot 2000).

Labour's approach is an archetypal liberal solution, which 'recognises inequity but seeks to cure that inequity within an existing set of social mechanisms' (Harvey 1973: 136). It implies a distributive theory of justice, which focuses on access to material goods as a means of need fulfilment. One might critique Labour's current water policies by drawing on current critiques of the distributive paradigm that advocate enablement (Gleeson 1997), or an equalization of capabilities to function rather than mere access to material goods (Sen 1999). This echoes the presupposition of the nationalized era, supplying a service rather a commodity, focusing on the derived demand (for example public health benefits, hygiene) rather than the material good. Labour's moral stance, it should be noted, is consistent with the distributive paradigm and inconsistent with the alternative notions of justice referred to above. It resuscitates a notion of justice-as-virtue, rather than the narrower Thatcherite conception of justice-as-freedom. But both conceptions of justice as enacted through the welfare state 'construct citizens as client-consumers, discouraging their active participation in public life . . . [and] the distributive paradigm of justice functions ideologically to reinforce this depoliticization' (Young 1990: 66). Interest-group pluralism, rather than a broader-based democratization of water policy (which might take the form of a reinclusion of local government and greater consumer representation on water company boards), has generally characterized the post-privatization policy process. As Young (1990) notes, the well-meaning 'ethic of care' advocated by some theorists of social justice functions to restrict public conflict to issues of distribution, not equally important issues of exclusion, at odds with Labour's stated concern over social exclusion in Britain.

Reflections on a political ecology of water

What does a political ecological approach contribute to the debate? First, this perspective may enable a more nuanced analysis of the 'winners' and 'losers' of privatization, and of the state's redistributive role. Detailed understanding of the implications of commodification in particular historical-geographical contexts requires analyses of specific modes of regulation of specific resources. These quotidian practices of regulation develop within, reinforce, and sometimes undermine broader macroeconomic patterns of resource regulation. This definition implies the articulation of institutional scales, and requires the treatment of regulation as both economic imperative and social practice (Bakker 2000). More profoundly, it requires an analysis of the mutually constitutive interrelationships between the discursive, social, and material dimensions of environmental change and socio-economic restructuring. Thus arises the tendency for political ecological work to bridge scales while examining the construction of those scales. Ironically, perhaps, for a research tradition that engages explicitly with the non-human, much political ecological work employs ethnographic methodologies, confronting the issue of agency—of both humans and non-humans—in a way that much political economic research does not.

Second, in beginning from the assertion that political economic analyses must incorporate not some idealized 'nature', but rather analyse specific resources in all of their particularities and complexities, a political ecological perspective can generate useful insights into the process of resource commodification. By analysing adaptive as well as transformative labour processes, for example, we can better identify and explain the various types and degrees of barriers to commodifying and capturing differential rents from different resources. In other words, a political ecological perspective can help us retheorize resource regulation and understand why water privatization is re-emerging at the turn of the twenty-first century despite the fact that water remains a liminal resource for capitalism.This may in turn contribute to a more sophisticated articulation of alternatives when confronted with a menu of privatization and private-public partnership options which treats water as being little different from, say, electricity, telecommunications, or roads.

Finally, political ecology analyses the impacts of water privatization and commercialization rather differently than, perhaps, a strictly political-economic approach. This is because political ecology wrestles simultaneously with questions of social justice and environmental justice. Political ecology not only begins from the assumption that socio-economic and environmental change are co-produced, but also broadens the set of actors—non-humans, as well as humans—who are considered both as objects of study, and also as holders of legitimate claims to equitable treatment.

Market environmentalism: who wins, who loses?

Underlying the dramatic shift in Labour's policy approach towards water in the 1990s is a more generalized transformation in the production of public goods in industrialized societies. Water is a liminal, yet also a highly strategic resource for capital accumulation. This strategic role remains a key justification for public involvement in water management, and was a key factor in the increase of public control of water supply and management (particularly in OECD countries) over the twentieth century, a process in which water resources and often supply networks were taken under public ownership, and water was supplied at subsidized rates, allocated by the state via public policy mechanisms rather than the market. Access to a water service was regarded as a precondition of participation in collective social activity; an entitlement, extended to all. Access to water supply is a material emblem of citizenship, and thus 'the withholding of water is a symbolic denial of the essence of "human being"... the legislation which prevents this not only provides for basic needs, it also upholds democratic participation in society' (Strang 2001: 61). Water was considered to be a public good, not one which could be sold like a commodity, or from which profits should be made.

The gradual commercialization of water management, and privatization of the water companies, was predicated upon the belief that water was—or should be—a tradable good rather than a public good. By the mid-twentieth century, the category of public goods had been extended to include many of the basic goods upon which citizens of industrialized countries depend for daily subsistence.[4] Privatization formalized an important transformation in the underlying conceptions of justice in the welfare state, through dismantling what Walzer (1983) would term the separate 'sphere of justice' for utility network services and reclassifying their products as commodities.

One important implication has been the resulting transformation in the entitlements of both humans and non-humans. The environmental implications of our production and consumption of water were in the past largely excluded from the regulatory framework, to which the entitlements of consumers and labour were central. Post-privatization, the environment is a legitimate user whose needs and protection have received increasing priority. Environmental groups have seen a considerable increase in their influence in the policy-making process; in contrast to organized labour. Environmental externalities are addressed within the water policy framework; water quality and environmental protection have with few exceptions improved since 1989. The environment, rather than the consumer, is increasingly the key driver of capital expenditure in the water industry. In contrast, social externalities are now, to a greater degree than in the past, excluded from the water policy framework.

Whereas the social costs of water production were previously externalized from the sphere of the politicised citizen to the environment, the environmental costs of water production are now externalized from the sphere of capitalized environment to consumers. Of course, the separation between 'environmental' costs and 'social' costs and the redesignation of entitlement-holders as 'internal' and 'external' to the workings of the state–capital nexus is entirely artificial; as David Harvey has repeatedly pointed out, ecological projects are always socio-political projects (and vice versa) (Harvey 1996). The task for the analyst, and that attempted in this book, is to understand 'who wins, and who loses', and to identify how our collective commitment to social and environmental justice has been fundamentally altered as a result. As argued in the introductory chapters, the public/private binary holds little analytical purchase when addressing these questions. Attention should rather be given to processes of resource regulation. The emergence of modes of resources regulation that facilitate the rational and sustainable use of nature, is highly functional—indeed necessary—to continued capital accumulation, and, is frequently accelerated and legitimated by political crises over resource management. The commercialization of water enacts a qualitative change in our relationship with the environment, increasingly viewed as a legitimate user of water whose interests are to be protected. But the commercialization of

---

[4] Economists' definitions of public goods have evolved over the twentieth century. Samuelson's classic definition referred to public goods as 'collective consumption goods' (Samuelson 1954). Subsequent definitions have defined public goods in terms of non-excludability and non-rivalry in consumption, and current definitions often acknowledge the existence of 'pure' public goods, 'impure' or 'mixed' public goods, and private goods. A classic example of a public good, under both definitions, is national defence. Economists dispute to what degree water supply is a public good, as the public health dimensions of water consumption and sewerage provision are more of a 'pure' public good than the water supply itself (Panayotou 1993; Pickart 2001).

water also reconfigures and in some cases constrains entitlements—of both humans and non-humans. Privatization and commercialization reallocate the costs and benefits of production, producing winners as well as losers. For a resource as essential to life and as emotive as water, this process will never unfold uncontested, for the re-regulation of water supply reconfigures not only our political economies, but also our relationships with the environment, and with one another.

BIBLIOGRAPHY

Akrich, M. (1992), 'The De-Scription of Technical Objects', in W. E. Bijker and J. Law (eds.), *Shaping Technology: Building Society, Studies in Socio-technical Change* (Boston: MIT Press), 205–24.

Aldred, J. (1994), 'Existence Value, Welfare and Altruism', *Environmental Values*, 3: 381–402.

Allan, J. A. and Karshenas, M. A. (1996), 'Managing Environmental Capital: The Case of Water in Israel, Jordan, the West Bank and Gaza, 1947–1995', in J. A. Allan (ed.), *Water, Peace and the Middle East: Negotiating Resources in the Jordan Basin* (Oxford: Tauris Academic Publishers), 75–119.

Altvater, E. (1989), 'Ecological and Economic Modalities of Time and Space', *Capitalism, Nature, Socialism*, 3: 59–71.

—— (1993), *The Future of the Market: An Essay on the Regulation of Money and Nature After the Collapse of 'Actually-existing Socialism'* (London: Verso).

AMA (1993), *Water, Warmth and Anti-Poverty Strategy* (London: Association of Metropolitan Authorities).

Anderson, T. L. and Leal, D. (1991), *Free Market Environmentalism* (Boulder, Colo.: Westview Press).

Anon. (1996), 'Will Mergers Boom Continue in 1996?' *Water Briefing*, 24 Jan. 10.

Armstrong, M., Cowan, S., and Vickers, J. (1994), *Regulatory Reform: Economic Analysis and the British Experience* (London and Cambridge Mass.: MIT Press).

Arnell, N. W., Jenkins A., and George, D. (1994), *The Implications of Climate Change for the National Rivers Authority* (Bristol: NRA).

Ashworth, W. (1960), *An Economic History of England: 1870–1939* (London: Methuen).

Averch, H. and Johnson, L. (1962), 'Behaviour of the Firm Under Regulatory Constraints', *American Economic Review*, 52: 1052–69.

Bakker, K. (1999*a*), 'Privatising the Environment: A Political Ecology of Water Privatization in England and Wales, 1989–1999', unpublished D.Phil. Thesis, Department of Geography, Oxford University.

—— (1999*b*), 'Deconstructing Discourses of Drought', *Transactions of the Institute of British Geographers*, 24: 367–78.

—— (2000), 'Privatising Water, Producing Scarcity: The Yorkshire Drought of 1995', *Economic Geography*, 76/1: 4–27.

—— (2001), 'Paying for Water: Water Charging and Equity in England and Wales', *Transactions of the Institute of British Geographers*, 2 (June), 143–64.

—— (2002), 'From State to Market: Water *mercantilización* in Spain', *Environment and Planning A*, 34: 767–90.

Bannister, N. (1998), 'Pre-paid Water Meters Banned', *Guardian*, 21 Feb., 22.

Barkas, E. and Walker, C. (1996), 'Yorkshire Water Faces £150m. Bill for Drought', *Yorkshire Post*, 29 Feb.

Barraqué, B. (2001), 'De l'appropriation à l'usage : l'eau, patrimoine commun', in M. Caron and J. Fromageau (eds.), *Genèse du droit de l'environnement, ii. Droit des espaces naturels et des pollutions* (Paris: L'Harmattan), 213–39.

Bateman, I., Langford, I., Nishikawa, N., and Lake, I. (2000), 'The Axford Debate Revisited: A Case Study Illustrating Different Approaches to the Aggregation of Benefits Data, *Journal of Environmental Planning and Management*, 43/2: 291–302.

Bauer, J. M. (1995), 'The Regulatory Treatment of Utility Diversification', *Land Economics,* 71/3: 386–400.

BBC (1998), 'Wessex Water Falls to US Bid', *BBC News Online*, www.news.bbc.co.uk. 6 Aug.

Beck, U. (1992), *Risk Society: Towards a New Modernity* (London: Sage).

Benton, T. (1989), 'Marxism and Natural Limits: An Ecological Critique and Reconstruction', in *The Greening of Marxism* (London and New York: Guildford, 157–83.

—— (1996), *The Greening of Marxism* (London and New York: Guildford).

Bilsborough, S. (1997), 'Pricing the Countryside: The Example of Tir Cymen', in J. Foster (ed.), *Valuing Nature? Economics, Ethics and the Environment* (London: Routledge), 89–102.

Birchall, J. (ed.) (2001), *The New Mutualism in Public Policy* (London: Routledge).

—— (2002), 'Mutual, Non-profit, or Public Interest Company? An Evaluation of Options for the Ownership and Control of Water Utilities', *Annals of Public and Cooperative Economics*, 73/2: 181–213.

Black, John (1997), *Oxford Dictionary of Economics* (Oxford: Oxford University Press).

Blokland, M., Braadbaart, O., and Schwartz, K., (1999), *Private Business, Public Owners: Government Shareholders in Water Enterprises* (Nieuwegein, Neths.: Ministry of Housing, Spatial Planning, and Environment.

BMA (1994), *Water: A Vital Resource* (London: British Medical Association).

Boardman, B. (1991), *Fuel Poverty* (London: Belhaven).

—— (1999), *Equity and the Environment* (London: Friends of the Earth).

Bond, P. (2001), *Against Global Apartheid: South Africa Meets the World Bank, IMF and International Finance* (Cape Town: University of Cape Town Press).

Bond, S., Klemm, A., and Simpson, H. (2001), *Labour and Business Taxes*, Election briefing note no. 6 (London: Institute for Fiscal Studies).

Boulton, L. (1996), 'Yorkshire Water in £70m Drought Fight', *Financial Times*, 27 April, 4.

Bowers, J. (1993), 'A Conspectus on Valuing the Environment', *Journal of Environmental Planning and Management*, 36/1: 91–100.

British Water (1995), *British Water: Expertise Worldwide. Annual Report 1993–1995* (London: British Water).

—— (1998), *British Water: Annual Report 1997/98* (London: British Water).

Brown, A. (1992*a*), 'Inland Water Quality and Pollution', in A. Brown (ed.), *The UK Environment* (London: HMSO), 87–105.

—— (1992*b*), *The UK Environment* (London: Department of the Environment).

Buckley, C. (1996), 'Yorkshire Water Counts the Cost of Supply Hitches', *The Times*, 29 Feb.

Buer, M. V. (1926), *Health, Wealth and Population in the Early Days of the Industrial Revolution* (London: Routledge).

Buller, H. (1996), 'Privatization and Europeanization: The Changing Context of Water Supply in Britain and France', *Journal of Environmental Planning and Management*, 39/4: 461–82.

Bunyan, N. (1995), 'Drought Firm Plans Second £50m Pipeline', *Yorkshire Post*, 15 Dec.

Burgess, J., Harrison C., and Clark, J. (1998), 'Respondents' Evaluation of a CV Survey: A Case Study Based on an Economic Valuation of the Wildlife Enhancement Scheme, Pevensey Levels in East Sussex', *Area*, 30/1: 19–27.

Burke, T. (1996), *Lifebuoy Men, Lux Women: Commodification, Consumption and Cleanliness in Modern Zimbabwe* (London: Duke University Press).

Burkett, P. (1999), *Marx and Nature: A Red-Green Perspective* (New York: St Martin's Press).

Burt, T. and Haycock N. E. (1992), 'Catchment Planning and the Nitrate Issue: A UK Perspective', *Progress in Physical Geography*, 16/4: 379–404.

—— Heathwaite A., and Trudgill, S. (eds.) (1993), *Nitrate: Processes, Patterns and Management* (Chichester: Wiley).

Cairncross, A. (1992), *The British Economy Since 1945: Economic Policy and Perforamnce, 1945–1990* (Oxford: Blackwell).

Castree, N. (1995), 'The Nature of Produced Nature: Materiality and Knowledge Construction in Marxism', *Antipode*, 27/1: 12–48.

—— and Braun B. (1998), 'The Construction of Nature and the Nature of Construction: Analytical and Political Tools for Building Survivable Futures', in B. Braun and N. Castree (eds.), *Remaking Reality: Nature at the Millenium* (London: Routledge), 3–42.

Chadwick, E. (1842), *Report on the Sanitary Condition of the Labouring Population of Great Britain* (London: HMSO).

Chick, M. (1998), *Industrial Policy in Britain 1945–1951: Economic Planning, Nationalisation, and the Labour Governments* (Cambridge: Cambridge University Press).

Christoff, P. (1996), 'Ecological Modernisation, Ecological Modernities', *Environmental Politics*, 5/3: 476–500.

CIPFA (1979), *Water Supply and Sewage Treatment and Disposal Statistics, 78–79* (London: CIPFA).

—— (1987), *The Water Industry Charges for Services in 1986/87* (London: CIPFA).

Clark, G. (1992a), 'Commentary: "Real" Regulation', *Environment and Planning A*, 24: 613–14.

—— (1992b), '"Real" Regulation: The Administrative State', *Environment and Planning A*, 24: 615–27.

Clarke, T. (1993), 'The Political Economy of the UK Privatization Programme: A Blueprint for Other Countries?' in T. Clarke and C. Pitelis (eds.), *The Political Economy of Privatization* (London and New York: Routledge).

Coase, R. H. (1960), 'The Problem of Social Cost', *Journal of Law and Economics*, 3: 1–44.

Conrad, J. (1990), *Nitrate Pollution and Politics* (Aldershot: Avebury).

Consumer's Association (2000), *Water Competition: Pipe Dreams?* (London: Consumer's Association/Which).

Cowan, S. (1993), 'Regulation of Several Market Failures: The Water Industry in England and Wales', *Oxford Review of Economic Policy*, 9/4: 14–23.

—— (1994), 'Privatization and Regulation of the Water Industry in England and Wales', in M. Bishop, J. Kay, C. Mayer, and Thompson D. (eds.), *Privatization and Regulation: The UK Experience* (Oxford: Oxford University Press), 112–36.

—— (1997a), 'Competition in the Water Industry', *Oxford Review of Economic Policy*, 13/1: 83–92.

—— (1997b), 'Market and Regulatory Failure in the Water Sector', in U. Collier (ed.), *Deregulation in the European Union: Environmental Perspectives* (London: Routledge), 131–44.

Cryer, R. (1995), 'Changing Responses to Water Resource Problems in England and Wales', *Geography*, 80/1: 45–57.

Curwen, P. (1994), *Privatization in the UK: Facts and Figure* (London: Ernst & Young).

Dale, B. (1995), 'The Investor's Judgement', in CIPFA (ed.), *The Water Industry: Looking Forward from the Periodic Review* (London: CIPFA), 27–40.

Darier, E. (1999), *Discourses of the Environment* (Oxford: Blackwell).

Dasgupta, P. (1990), 'The Environment as a Commodity', *Oxford Review of Economic Policy*, 6/1: 51–67.

Day, D. (1987), 'An Australian Perspective on Drought and Water Management Objectives for Regional Development', *Water Resources Development*, 3/4: 266–83.

Defeuilley, C. (1998), 'Régulation et service public: Enseignements tires de l'expérience britannique', *Flux*, 31/32: 49–60.

DEFRA (2001a), 'UK Maintains Record-breaking Performance for River Quality', News Release, 5 Nov. (London: Department for Environment, Food and Rural Affairs).

—— (2001b), *Tuning Water Taking: Government Decisions Following Consultation on the Use of Economic Instruments in Relation to Water Abstraction* (London: Department for Environment, Food and Rural Affairs).

—— (2002), 'A Better Deal and Continuing Protection for Water Customers', News release, 19 Mar. (London: Department for Environment, Food and Rural Affairs).

Demand Management Centre (1994), 'NRA Launches its National Water Resources Strategy' *Demand Management Bulletin*, 1.

Den Hartogh, G. (1999), 'The Architectonic of Michael Walzer's Theory of Justice', *Political Theory*, 27/4: 491–522.

DETR (1998a), *Water Charging in England and Wales: A New Approach*, Consultation paper (London: Department of the Environment, Transport and Regions).

—— (1998b), *Water Charging in England and Wales: Government decisions Following Consultation* (London: Department of the Environment, Transport and Regions).

—— (1998c), *The Review of the Water Abstraction Licensing System in England and Wales*, Consultation Paper (London: Department of the Environment, Transport and Regions).

—— (1999a), *Taking Water Responsibly: Government Decisions Following Consultation on Changes to the Water Abstraction Licensing System in England and Wales* (London: Department of the Environment, Transport and Regions).

—— (1999b), *Water Industry Act 1999: Consultation on Regulations* (London: Department of the Environment, Transport and Regions).

—— (2000a), *Competition in the Water Industry* (London: Department of the Environment, Transport and Regions).

—— (2000b), *Economic Instruments in Relation to Water Abstraction* (London: Department of the Environment, Transport and Regions).

—— (2000c), *Water Industry Act 1999: Delivering the Government's Objectives* (London: Department of the Environment, Transport and Regions).

DFID (1998a), *Better Water Services in Developing Countries: Public–Private Partnership—the Way Ahead* (London: Department for International Development).
—— (1998b), *Water Matters: DFID's Role in Water for Development* (London: Department for International Development).
Dimcock, M. E. (1933), *British Public Utilities and National Development* (London: Allen & Unwin).
Dobson F. (1995a), *Money Down the Drain: Water Leakage and Water Metering* (London: Labour Party).
—— (1995b), *More Money Down the Drain: Water Leakage and Water Profits* (London: Labour Party).
DOE (1976), *Review of the Water Industry in England and Wales* (London: Department of the Environment).
—— (1979), *Water Data* (London: Department of the Environment).
—— (1986), *Privatisation of the Water Authorities in England and Wales* (London: Department of the Environment).
—— (1992), *Using Water Wisely* (London: Department of the Environment).
—— (1995), *Water Conservation: Government Action* (London: Department of the Environment and the Welsh Office).
—— (1996a), *The Environment Agency and Sustainable Development* (London: HMSO).
—— (1996b), *Review of the Potential Effects of Climate Change in the United Kingdom* (London: HMSO).
—— (1996c), *Increasing Customer Choice: Competition in the Water and Sewerage Industries: The Government's Proposals* (London: HMSO).
—— (1996d), *Water Resources and Supply: Agenda for Action* (London: HMSO).
Downing, T., and Bakker, K. (2000), 'Drought Discourse and Vulnerability', in D. A. Wilhite (ed.), *Drought, Hazards and Disasters: A Series of Definitive Major Works* (London: Routledge).
Dracup, S. B. (1973a), 'Water Supply in Great Britain 1690–1950: A Brief History in Six Parts', *British Water Supply* (Jan.), 17–19.
—— (1973b), 'Water Supply in Great Britain 1690–1950: A Brief History in Six Parts', *British Water Supply* (Feb.), 15–19.
—— (1973c), 'Water Supply in Great Britain 1690–1950: A Brief History in Six Parts', *British Water Supply* (Mar.), 20–3.
—— (1973d), 'Water Supply in Great Britain 1690–1950: A Brief History in Six Parts', *British Water Supply* (Apr.), 24–7.
—— (1973e), 'Water Supply in Great Britain 1690–1950: A Brief History in Six Parts', *British Water Supply* (May), 26–32.
Drakeford, D. (1995), *Token Gesture: A Report on the Use of Token Meters by the Gas, Electricity and Water Companies* (Cardiff: National Local Government Forum Against Poverty).
Drakeford, M. (1997), 'The Poverty of Privatization: Poorest Customers of the Privatized Gas, Water and Electricity Industries', *Critical Social Policy*, 17: 115–32.
Drummond, I. and Marsden, T. K. (1995), 'Regulating Sustainable Development', *Global Environmental Change*, 5: 51–64.
DTI (1998a), *A Fair Deal for Consumers: Modernising the Framework for Utility Regulation* (London: Department of Trade and Industry).

—— (1998*b*), *A Fair Deal for Consumers: Modernising the Framework for Utility Regulation: Response to Consultation* (London: Department of Trade and Industry).

Du Gay, P. and Salaman, G. (1992), 'The Cult[ure] of the Customer', *Journal of Management Studies*, 29 /5: 615–33.

DWI (2000), *Drinking Water 2000: A Report by the Chief Inspector* (London: Drinking Water Inspectorate).

EA (1996*a*), *Introductory Guidance on the Agency's Contribution to Sustainable Development* (Bristol: Environment Agency).

—— (1996*b*), *The Environment Agency: Management Statement* (London: Department of the Environment).

—— (1996*c*), *Review of Water Company Plans to Safeguard Summer Water Supplies: Environment Agency Report to the Secretary of State for the Environment* (Bristol: Environment Agency).

—— (1996*d*), *Interim Report on the Environmental Impacts of the Drought on Yorkshire's Rivers: April 1995–April 1996* (Leeds: Environment Agency).

—— (1998*a*), *Water Resources Planning Guideline* (Bristol: Environment Agency).

—— (1998*b*), *The Environmental Costs and Benefits of Water Resources: A Preliminary Methodology* (Bristol: Environment Agency).

—— (1998*c*), *The Environment Agency's State of the Environment Report for Thames Region* (Reading: Environment Agency (Thames Region)).

—— (1998*d*), *Thames Environment 21: The Environment Agency Strategy for Land-use Planning in the Thames Region.* (Reading: Environment Agency).

—— (1998*e*), *Progress in Water Supply Planning: The Environment Agency's Review of Water Company Water Resource Plans* (Bristol: Environment Agency).

—— (2001),'Decade of Clean-up Brings best-ever River and Estuary Quality Results', Press release, 5 Nov. (Bristol: Environment Agency).

—— (2003), 'Spotlight on Business Performance' (Bristol: Environment Agency)

Eagleton, T. (1991), *Ideology* (London: Verso).

Eckersley, R. (1995), 'Markets, the State and the Environment: An Overview', in R. Eckersley (ed.), *Markets, The State, and The Environment: Towards Integration* (London: Macmillan), 7–45.

Emel, J. (1990), 'Resource Instrumentalism, Privatization and Commodification', *Urban Geography*, 11/6: 527–47.

—— and Brooks, E. (1988), 'Changes in Form and Function of Property Rights Institutions under Threatened Resource Scarcity', *Annals of the Association of American Geographers*, 78/2: 241–52.

—— and Roberts, R. (1995), 'Institutional Form and Its Effect on Environmental Change: The Case of Groundwater in the Southern High Plains', *Annals of the Association of American Geographers*, 85/4: 664–83.

—— —— and Sauri, D. (1992),'Ideology, Property, and Groundwater Resources: An Exploration of Relations, *Political Geography*, 11/1: 37–54.

Enzensberger, H. M. (1996), 'A Critique of Political Ecology', in T. Benton (ed.), *The Greening of Marxism.* (London and New York: Guildford), 17–49.

Ernst, J. (1994), *Whose Utility? The Social Impact of Public Utility Privatization and Regulation in Britain* (Milton Keynes: Open University Press).

Escobar, A. (1996), 'Constructing Nature: Elements for a Poststructural Political Ecology', in R. Peet and M. Watts (eds.), *Liberation Ecologies: Environment, Development and Social Movements* (London: Routledge), 46–68.

ESRI (1994), *Waste Water Services: Charging Industry the Capital Cost* (Dublin: The Economic and Social Research Institute).

Farrar, C. (1996), 'Yorkshire Water Stuns Customers with its Latest Letter: Because We've Been So Efficient, Here's a Cheque for £10', *Yorkshire Post*, 16 Sept.

Ferner, A. and Colling, T. (1993), 'Privatization of the British Utilities: Regulation, Decentralization and Industrial Relations', in T. Clarke and C. Pitells (eds.), *The Political Economy of Privatization* (London: Routledge), 125–41.

Fletcher, P. (2001a), 'Regulatory Developments: Moving Towards Total Competition for Utilities?' Talk to the Adam Smith Institute, 6th Annual Conference on the Future of Utilities, London, 14 Mar.

—— (2001b), 'Regulation, Competition and the Water Industry', Talk to the Utility Conference 2001, Brighton. 17 Oct.

—— (2002), 'Water Regulation: The Next Decade', Speech to the Institute of Economic Affairs Conference 'Water 2002', London, 11 June.

—— (2003), Speech to Water UK, City Conference, London, 15 Jan.

Foster, C. D. (1992), *Privatization, Public Ownership and the Regulation of Natural Monopoly* (Oxford: Blackwell).

Foster, J. B. (2000), *Marx's Ecology: Materialism and Nature* (New York: Monthly Review Press).

Franceys, R. (1997), *Private Sector Participation in the Water and Sanitation Sector*. DFID Water Resources Occasional Papers (Loughborough and London: Water and Engineering Development Centre and the Department for International Development).

Frankham, J. and Webb, M. (1977), 'The Principle of Equalization and Charging for Water', *Public Finance and Accountancy*, 4/6: 196–202.

Fraser, D. (1973), *The Evolution of the British Welfare State: A History of Social Policy Since the Industrial Revolution* (London: Macmillan).

Gandy, M. (1997), 'The Making of a Regulatory Crisis: Restructuring New York City's Water Supply', *Transactions of the Institute of British Geographers*, 22: 338–58.

—— (1999), 'The Paris Sewers and the Rationalization of Urban Space, *Transactions of the Institute of British Geographers*, 24/1: 23–44.

Garrett, P. (1996), 'Take Your Partners: Partnership not Takeover is the Best Way to Success, believes Thames Water's Managing Director Bill Alexander', *Utility Week*, 8 Mar., 20.

—— (2002), 'Fletcher: The Industry Needs a Water Bill', *Utility Week*, 11 Jan. 6.

Garrod, G. and Willis, K. (1994), *The Transferability of Environmental Benefits: A Review of Recent Research in Water Resources Management*, Working Paper Series (Newcastle: Centre for Rural Economy).

Getzler, J. (1993), 'Rules Writ in Water: A History of Riparian Rights and Property Use Doctrine in England to 1870', unpublished doctoral thesis, Department of History, Oxford University.

Gibbs, D. (1996), 'Integrating Sustainable Development and Economic Restructuring: A Role for Regulation Theory?' *Geoforum*, 27/1: 1–10.

Gibson-Graham, J. (1996), *The End of Capitalism (As We Knew It): A Feminist Critique of Political Economy* (Oxford:Blackwell).

Gill, A. (1995), 'Digging a Watery Grave', *Yorkshire Post*, Leeds.

Gleeson, B. (1997), 'Community Care and Disability: The Limits to Justice', *Progress in Human Geography*, 212: 199–224.

Gleick, P. (ed.) (1993), *Water in Crisis: A Guide to the World's Fresh Water Resources* (Oxford: Oxford University Press).
—— (2000), 'The Changing Water Paradigm: A Look at Twenty-first Century Water Resources Development', *Water International*, 25/1: 127–38.
Global Water Intelligence (2001), 'UK: A Lacklustre Results Season', *Global Water Intelligence*. Source: htpp://www.globalwaterintel.com/Samples/july01.htm. Accessed 22/10/01.
Global Water Report (1996), 'Anglia, Yorkshire Temper Foreign Enthusiasm', *UK Briefing* (4 Aug.), 3.
—— (1997), 'Competition's Champion', *UK Briefing*, 18 (Mar.), 16–18.
Glynn, D. (1992), 'Long Run Regulation of Natural Monopoly: The Case of the Water Industry', in T. Gilland (ed.), *Incentive Regulation* (London: Centre for the Study of Regulated Industry).
Goubert, J. P. (1986), *The Conquest of Water* (London: Polity Press).
Graham, S. (1997), 'Liberalized Utilities, New Technologies, and Uurban Social Polarization: The UK Experience', *European Urban and Regional Studies*, 4/2: 135–50.
—— and S. Marvin (1994), 'Cherry Picking and Social Dumping: Utilities in the 1990s', *Utilities Policy*, 4/2: 113–20.
—— —— (1995), 'More than Ducts and Wires: Post-Fordism, Cities and Utility Networks', in P. Healey, S. Cameron, S. Davodi, S. Graham and A. Madhani-Pour (eds.), *Managing Cities: The New Urban Context* (London: John Wiley & Sons), 169–80.
—— —— (2001), *Splintering Urbanism: Networked Infrastructures, Technological Mobilities and the Urban Condition* (London: Routledge).
Gregg, P. (1950), *A Social and Economic History of Britain: 1760–1972* (London: Harrap).
Griffiths, E. (1976), 'A Conservative View of the Review', *Water: The Journal of the National Water Council* (Sept.), 2–4.
Grove-White, R. (1997), 'The Environmental "Valuation" Controversy: Observations on Its Recent History and Significance', in J. Foster (ed.), *Valuing Nature? Economics, Ethics and the Environment* (London: Routledge), 21–31.
Guy, S. and Marvin, S. (1996*a*), 'Managing Water Stress: The Logic of Demand-Side Infrastructure Planning', *Journal of Environmental Planning and Management*, 39/1: 123–8.
—— —— (1996*b*), 'Transforming Urban Infrastructure Provision: The Emerging Logic of Demand-Side Management', *Policy Studies*, 17/2: 137–47.
—— Graham, *et al.* (1997), 'Splintering Networks: Cities and Technical Networks in 1990s Britain', *Urban Studies*, 34/2: 191–216.
Haddon, M. (1997), 'The Dam Buster', *Utility Week*, 25 Apr., 12–13.
Hajer, M. (1995), *The Politics of Environmental Discourse: Ecological Modernisation and the Policy Process* (Oxford: Clarendon Press).
—— (1996), 'Ecological Modernization as Cultural Politics', In S. Lash, B. Szerszynski, and B. Wynne (eds.), *Risk, Environment and Modernity* (London: Sage), 246–68.
Hall, D. (1996), 'Economic Instruments to Mitigate Water Scarcity: Marginal Cost Rate Design and Wholesale Water Markets', in D. Hall (ed.), *Advances in the Economics of Environmental Resources* (Greenwich, Conn. and London: JAI Press).
Hall, D. and Lobina, E. (2001), *Private to Public: International Lessons of Water Remunicipalisation in Grenoble, France* (University of Greenwich: Public Services International Research Unit).

Hannah, L. (1992), 'The Economic Consequences of the State Ownership of Industry', in R. Floud and D. McCloskey (eds.), *The Economic History of Britain since 1700* (Cambridge: Cambridge University Press), i. 168–94.

Hardt, M. and Negri, A. (2000), *Empire* (Cambridge, Mass. and London: Harvard University Press).

Harper, W. R. (1988), 'Privatization in the Water Sector', in V. V. Ramanadhan (ed.), *Privatization in the UK* (London: Routledge), 215–25.

Harrison, D. (1996), 'Poor Hit by Secret Water Cut-Offs', *Observer*, 25 Aug., 3.

Harvey, D. (1973), *Social Justice and the City* (London: Edward Arnold).

—— (1979), 'Population, Resources, and the Ideology of Science'. in S. Gale and G. Olsson (eds.), *Philosophy in Geography* (London: D. Reidel), 155–85.

—— (1982), *The Limits to Capital* (Oxford: Blackwell).

—— (1985), 'The Geopolitics of Capitalism', in D. Gregory and J. Urry (eds.), *Social Relations and Spatial Structure* (London: Macmillan), 128–63.

—— (1996), *Justice, Nature and the Geography of Difference* (Oxford: Blackwell).

Hassan, J. (1985), 'The Growth and Impact of the British Water Industry in the Nineteenth Century', *Economic History Review*, 38/4: 531–47.

—— (1993), 'The Water Industry 1900–51: A Failure of Public Policy?' Discussion Papers in Economics and Economic History. Manchester Metropolitan University (Series 93: 09).

—— (1995), 'The Impact of EU Environmental Policy on Water Industry Reform', *European Environment*, 5: 45–51.

—— (1998), *A History of Water in Modern England and Wales* (Manchester: Manchester University Press).

—— and Taylor, P. (1996), *The Politics of Water in Early and Mid-Victorian Britain* (Manchester: Manchester Metropolitan University).

Haughton, G. (1998), 'Private Profits—Public Drought: The Creation of a Crisis in Water Management for West Yorkshire', *Transactions of the Institute of British Geographers*, 23/4: 419–34.

Hawken, P. and Lovins, A. (1999), *Natural Capitalism: Creating the Next Industrial Revolution* (Boston: Little, Brown).

Hay, C. (1994a), 'The Structural and Ideological Contradictions of Britain's Post-War Reconstruction', *Capital and Class*, 54: 25–59.

—— (1994b), 'Environmental Security and State Legitimacy', in M. O'Connor (ed.), *Is Capitalism Sustainable? Political Economy and the Politics of Ecology* (London: Guilford), 217–31.

—— (1995), 'Re-stating the Problem of Regulation and Re-regulating the Local State', *Economy and Society*, 24/3: 387–407.

—— (1996), *Re-Stating Social and Political Change*. (Milton Keynes: Open University Press).

Helm, D. (1994), 'British Utility Regulation: Theory, Practice and Reform', *Oxford Review of Economic Policy*, 10/3: 17–39.

—— and Jenkinson, T. (1997), 'The Assessment: Introducing Competition into Regulated Industries', *Oxford Review of Economic Policy*, 13/1: 1–14.

—— and Pearce, D. (1992), 'Assessment: Economic Policy Towards the Environment', *Oxford Review of Economic Policy*, 6/1: 1–16.

—— and Rajah, N. (1994), 'Water Regulation: The Periodic Review', *Fiscal Studies*, 15/2.

—— and Yarrow, G. (1988), 'The Assessment: The Regulation of Utilities', *Oxford Review of Economic Policy*, 4/2: pp. i–xxxi.

Herbert, A. and Kempson, E. (1995), *Water Debt and Disconnection* (London: Policy Studies Institute).

Herrington, P. (1996*a*), *Climate Change and the Demand for Water* (London: Department of the Environment).

—— (1996*b*), 'Pricing Water Properly', Discussion Papers in Public Sector Economics, 96/6 (Public Sector Economics Research Centre, University of Leicester).

Hirst, P. and Thomson, G. (1996), *Globalization in Question: The International Economy and the Possibilities of Governance* (Cambridge: Polity Press).

Holtham, G. (1997), 'The Water Industry: Why it Should Adopt the Mutual Society Model', *Journal of Co-operative Studies*, 29/3: 3–8.

Hopkinson, P., James, P., and Sammut, A. (2000), 'Environmental Performance Evaluation in the Water Industry of England and Wales', *Journal of Environmental Planning and Management*, 43/6: 873–95.

House of Commons Environment Committee (1996), *Water Conservation and Supply: Interim Report* (London: HMSO).

House of Commons Environmental Audit Committee (2000), *Water Prices and the Environment*, Seventh Report of the Select Committee on Environmental Audit.

Huby M. (1995), 'Water Poverty and Social Policy: A Review of Issues for Research', *Journal of Social Policy*, 24/2: 219–36.

—— and Anthony, K. (1997) 'Regional Inequalities in Paying for Water', *Policy Studies*, 18/3/4: 207–17.

Hughes, S. (1996), *Report for the Yorkshire Wildlife Trust on the Independent Commission of Inquiry into Water Supply in Yorkshire* (York: Yorkshire Wildlife Trust).

Hunt, L. and Lynk, E. (1995), 'Privatization and Efficiency in the UK Water Industry: An Empirical Analysis', *Oxford Bulletin of Economics and Statistics*, 57/3: 371–88.

Hunter, C., Rowe, I. *et al*. (1996), *Supply Augmentation or Demand-management? Two Case Studies of the Challenges of Meeting Water Demands in South-East England*, CUDEM (Leeds: Leeds Metropolitan University).

Illich, I. (1986), *H2O and the Waters of Forgetfulness* (London: Marion Boyars Publishers).

IOH (1995), *Hydrological Data United Kingdom: 1994 Yearbook* (Wallingford, Oxon.: Institute of Hydrology and British Geological Survey).

Jackson, P. M. (1997), 'The Privatization of the British Public Sector: An Assessment of a Policy Innovation', in M. Baldassarri, Macchiati, A., and D. Piacentino (eds.), *The Privatization of Public Utilities: The Case of Italy* (London: Macmillan), 73–98.

Jacobs, M. (1994), 'The Limits to Neo-Classicism: Towards an Institutional Environmental Economics', in M. Redclift and T. Benton (eds.) *Social Theory and the Global Environment* (London: Routledge), 67–91.

—— (1997), 'Environmental Valuation, Deliberative Democracy, and Public Decision-Making Institutions', in J. Foster (ed.), *Valuing Nature? Economics, Ethics and the Environment* (London: Routledge) 211–31.

—— (1990), 'Regulation Theories in Retrospect and Prospect', *Economy and Society*, 19/2: 153–216.

—— (1995), 'The Regulation Approach, Governance, and Post-Fordism', *Economy and Society*, 24/3: 307–33.

Jordan, A. G., Richardson, J. J. et al. (1977), 'The Origins of the Water Act 1973', *Public Administration*, 55: 317–34.
Kay, J. (1996a), 'The Future of UK Utility Regulation', in M. E. Beesley (ed.), *Regulating Utilities: A Time for Change?* (London: Institute for Economic Affairs), 145–71.
—— (1996b), 'Regulating Private Utilities: The Customer Corporation', *Journal of Co-operative Studies*, 29/2: 28–46.
Katz, C. (1998), 'Whose Nature, Whose Culture? Private Production of Space and the Preservation of Nature', in B. Braun and N. Castree (eds.), *Remaking Reality: Nature at the Millenium* (London: Routledge), 46–63.
Kennedy, W. and Delargy, R. (2000), *Explaining Victorian Entrepreneurship: A Cultural Problem? A Market Problem? No Problem?* Working Papers in Economic History, WP 61, Department of Economic History, LSE.
King, T. (1980), 'Learning Financial Discipline', *Water: The Journal of the National Water Council* (May), 2–3.
Kinnersley, D. (1988), *Troubled Water: Rivers, Politics and Pollution* (London: Hilary Shipman).
—— (ed.) (1992), *Privatization and the Water Environment in England and Wales*, Country Experiences with Water Resources Management (Washington: World Bank).
—— (1994), *Coming Clean: The Politics of Water and the Environment* (London: Penguin).
Labour Party (2002), 'Improving Health and Social Care', National Policy Forum Consultation document (London: Labour Party).
La Porte, T. R. (1994), 'Large Technical Systems, Institutional Surprises, and Challenges to Political Legitimacy', *Technology in Society*, 16/3: 269–88.
Lapsley, I. and Wright, H. (1990), 'On the Privatisation of British Rail', *Public Money and Management* (Autumn), 49–53.
Lascelles, D. and Martinson, J. (1996), 'Yorkshire Water Hit by £40m Revenue Penalty: Price Curb Agreed as Regulator Attacks "Serious Failures" ', *Financial Times*, 4 June, 22.
Latour, B. (1993), *We Have Never Been Modern* (New York and London: Harvester Wheatsheaf).
Lazenby, P. (1995), 'Water Stocks Continue to Fall as Cut-offs Loom', *Yorkshire Post*, 21 Sept.
—— (1996), 'Water Watchdog Anger as Profits Hit £162m', *Yorkshire Post*, 5 June.
Letza, S. and Smallman, C. (1996), 'Is Water Thicker than Blood? Overcoming the Conspiracy of Profit over Social Responsibility in Privatised Utilities', in L. Montanheiro, E. Rebelo, and G. Owen (eds.), *Public and Private Sector: Partnerships Working for Change* (Sheffield: Pavic Publications), 289–303.
Leys, C. (2001), *Market-driven Politics: Neoliberal Democracy and the Public Interest* (London: Verso).
Lingard, J. (1974), 'Pricing Water to Domestic Consumers', *National Westminster Bank Quarterly Review* (Feb.), 34–73.
Litfin, K. (1994), *Ozone Discourses: Science and Politics in Global Environmental Cooperation* (New York: Columbia University Press).
Littlechild, S. (1986), *Economic Regulation of Privatised Water Authorities* (London: Department of the Environment, HMSO).

—— (1988), 'Economic Regulation of Privatized Water Authorities and Some Further Reflections', *Oxford Review of Economic Policy*, 4/2: 40–68.
London Economics (1997), *Water Pricing: The Importance of Long-Run Marginal Costs* (London: Prepared for Ofwat by London Economics).
Luckin, B. (1986), *Pollution and Control: A Social History of the Thames in the Nineteenth Century* (Bristol: Adam Hilger).
McCully, P. (1996), *Silenced Rivers: The Ecology and Politics of Large Dams* (London: Zed Books).
McDowell, L. (1991), 'Valid Games? A Response to Erica Schoenberger', *Professional Geographer*, 44/2: 212–15.
McGuinness, T. and Thomas, D. (1997), 'The Diversification Strategies of the Privatised WaSCs in England and Wales: A Resource-based View', *Utility Policy*, 6/4: 325–39.
MacGuire, F. (1996), *The Environmental Threats of the Kielder Transfer Scheme* (London: Friends of the Earth).
Maclean, M. (1991), *French Enterprise and the Challenge of the British Water Industry: Water Without Frontiers* (Aldershot: Avebury Publishing Company).
McPherson, J. I. (1998), *Water Resources Act 1991/Environment Agency: Appeal by Thames Water Utilities Limited, Axford Abstraction* (London: Department of the Environment, Transport and Regions).
Maloney, W. A. and Richardson, J. (1994), 'Water Policy-Making in England and Wales: Policy Communities Under Pressure?', *Environmental Politics*, 3/4: 110–38.
—— —— (1995), *Managing Policy Change in Britain: The Politics of Water* (Edinburgh: Edinburgh University Press).
Marsh, D. (1991), 'Privatisation under Mrs. Thatcher: A Review of the Literature', *Public Administration*, 69: 459–80.
Marsh, T. (1995), 'Drought Returns to the United Kingdom', *Drought Network News*, 7/3: 5–6.
—— (1996), 'The (1995) UK Drought: A Signal of Climatic Instability?' *Proceedings of the Institute of Civil Engineers*, 118: 189–95.
—— and P. S. Turton (1996), 'The (1995) Drought: A Water Resources Perspective', *Weather* (Feb.), 47–53.
—— Monkhouse, R. A., Arnell, N. W., Lees, M. L., and Reynard, N. S. (1994), *The 1988–92 Drought* (Wallingford, Oxon.: Institute of Hydrology and British Geological Survey).
Martin, S. and Parker, D. (1997), 'Privatization: The Conceptual Framework', in S. Martin and D. Parker (eds.), *The Impact of Privatization: Ownership and Corporate Performance in the UK* (London: Routledge), 1–30.
Martín Mendiluce, J. M. (1996), 'Los embalses en España: Su necesidad y trascendencia económica', *Revista de Obras Públicas*, 3354: 7–24.
Martinson, J. (1996*a*), 'Companies and Finance: Yorkshire Water Moves to Restore its Reputation', *Financial Times*, 6 June.
—— (1996*b*), 'Companies and Finance: Yorkshire Hopes Tide has Turned', *Financial Times*, 24 July.
—— (1996*c*), 'Companies and Finance: Yorkshire Chief Faces Grilling', *Financial Times*, 26 July.
—— (1996*d*), 'Companies and Finance: Yorkshire Still "Very Keen" to Return Value', *Financial Times*, 30 Nov.

Martinson, J. (1996e), 'Tide Ebbs for Water Diversification', *Financial Times*, 29 Aug.
Marvin, S. (1994), 'Accessibility to Utility Networks: Local Policy Issues', *Local Government Studies*, 20/3: 437–57.
—— and Guy, S. (1997a), 'Consuming Water: Changing Logics of Water Management in Britain', *Journal of Urban Technology*, 4/3: 21–45.
—— —— (1997), 'Smart Metering Technologies and Privatised Utilities', *Local Economy* (August), 119–132.
—— Chappells, H. *et al.* (1998), *Pathways of Smart Metering Development: Shaping Environmental Innovation* (Newcastle: Centre for Urban Technology, Department of Town and Country Planning).
——, Graham, S., *et al.* (1996), *Privatised Utilities and Regional Management* (Harlow, Essex: South East Economic Development Strategy).
Marx, K. (1990, orig. 1867, i.), *Capital* (London: Penguin Classic).
Mayer, C. (2001), 'Water: The 1999 Price Review', in C. Robinson (ed.), *Regulating Utilities: New Issues, New Solutions* (London: Institute of Economic Affairs), 1–21.
Mayo, E. and Moore, H. (2001), *The Mutual State: How Local Communities Can Run Public Services* (London: New Economics Foundation).
Melville, A. (1994), 'Power, Strategy and Games: Economic Regulation of a Privatised Utility', *Public Administration*, 72/3: 385–408.
Meredith, S. (1992), 'Water Privatization: The Dangers and the Benefits', *Long Range Planning*, 25/4: 72–81.
Metropolitan Water Board (1949), *London's Water Supply* (London: Staples Press).
Middleton, J. and Saunders, P. (1997), 'Paying for Water', *Journal of Public Health Medicine*, 19/1: 106–15.
Miller-Bakewell, R. (1998), *Water: Crying Wolf*, Sector review (London: Global Securities Research and Economics Group, Merrill Lynch).
—— (2001a), *UK Water: Crossroads at Huntingdon*, Merrill Lynch In-depth Report (London: Merrill Lynch).
—— (2001b), *Water Services: The USA—An Untapped Market?*, Merrill Lynch In-depth Report (London: Merrill Lynch).
Millward, R. (1989), 'Privatization in Historical Perspective: The UK Water Industry', in D. Cobham, R. Harrington, and G. Zis (eds.), *Money, Trade and Payments: Essays in Honour of DJ Coppock* (Manchester: Manchester University Press), 188–209.
Ministerio de Medio Ambiente (1998), *Libro Blanco del Agua en España* (Madrid: MIMAM).
MMC (1981), *Report on the Severn-Trent Water Authority* (London: Monopolies and Mergers Commission).
Mol, A. (2002), Ecological Modernization and the Global Economy, *Global Environmental Politics*, 2/2: 92–115.
—— and Sonnenfeld, D. (2000), 'Ecological Modernization Around the World: An Introduction', *Environmental Politics*, 9/1: 3–16.
Monbiot, G. (2000), *Captive State: The Corporate Takeover of Britain* (London: Macmillan).
Moral Ituarte, L. and Saurí, D. (2000), 'Recent Developments in Spanish Water Policy: Alternatives and Conflicts at the End of the Hydraulic Era', *Geoforum*, 32: 351–62
Morse, L. (2000), 'A Case for Water Utilities as Cooperatives, and the UK Experience', *Annals of Public and Cooperative Economics*, 71/3: 467–93.

Mulholland, K. (2002), 'Throwing the Baby Out With the Bath-water: Managers and Managerialism in the Post-privatised Water Utilities, *Capital and Class*, 77 (Summer), 53–87.

National Audit Office (1998), 'Deficit in the Fund for Pensioners of the Former Water Authorities', Report by the Comptroller and Auditor General, HC 590 1997–98 (London: HMSO).

NCC (1993), *Paying the Price* (London: HMSO and National Consumer Council).

NEDO (1987), *Selling British Water Industry Expertise Overseas* (London: Civil Engineering Economic Development Committee on behalf of the National Economic Development Office).

Nevarez, L. (1996), 'Just Wait Until There's a Drought: Mediating Environmental Crises for Urban Growth', *Antipode*, 28/3: 246–72.

Newbery, D. (1999), *Privatization, Restructuring and Regulation of Network Utilities*, The Walras-Pareto Lectures (Cambridge, Mass.: MIT Press).

Newson, M. (1992), *Land, Water and Development: Sustainable Management of River Basin Systems* (London: Routledge).

Nixon, J. S. (1995), *Yorkshire Water Services Ltd: Applications for Emergency Drought Orders* (London: Department of the Environment).

Norgaard, R. B. (1985), 'Environmental Economics: An Evolutionary Critique and a Plea for Pluralism', *Journal of Environmental Economics and Management*, 12: 382–94.

NRA (1992a), *Water Resources Development Strategy: A Discussion Document* (Bristol: National Rivers Authority).

—— (1992b), *Managing the Drought and Water Resources: A Key Issues Statement* (Bristol: National Rivers Authority).

—— (1993a), *Water Resources Policy* (Bristol: National Rivers Authority).

—— (1993b), *Water Resources Strategy* (Bristol: National Rivers Authority).

—— (1993c), *Low Flows and Water Resources: Facts on the Top 40 Low Flow Rivers in England and Wales* (Bristol: National Rivers Authority).

—— (1994a), *Future Water Resources in the Thames Region: A Strategy for Sustainable Management* (Reading: National Rivers Authority (Thames Region)).

—— (1994b), *Water: Nature's Precious Resouce: An Environmentally Sustainable Water Resources Development Strategy for England and Wales* (London: National Rivers Authority).

—— (1994c), *Regional Water Resources Strategy: Northumbria and Yorkshire* (Leeds: National Rivers Authority,.

—— (1995a), *Saving Water: The NRA's Approach to Water Conservation and Demand Management* (Bristol: National Rivers Authority).

—— (1995b), *The Drought of 1995: A Report to the Secretary of State for the Environment* (Bristol: National Rivers Authority).

—— (1995c), *Measures to Safeguard Public Water Supplies: A Second Report to the Secretary of State for the Environment on the Drought of 1995* (Bristol: National Rivers Authority).

—— (1996a), *Refill Prospects: A Third Report to the Secretary of State for the Environment on the Drought of 1995/96* (Bristol: National Rivers Authority).

—— (1996b), *Review of Water Company Plans to Safeguard Water Supplies* (Bristol: National Rivers Authority).

O'Connell-Davidson, J. (1990), 'The Commercialisation of Employment Relations: The Case of the Water Industry', *Work, Employment and Society*, 4/4: 531–49.

O'Connell-Davidson, J. (1993), *Privatization and Employment Relations: The Case of the Water Industry* (London: Mansell).
O'Connor, J. (1996), 'The Second Contradiction of Capitalism', in T. Benton (ed.), *The Greening of Marxism* (London: Guilford Press), 197–221.
O'Connor, J. (1973), *The Fiscal Crisis of the State* (New York: St. Martin's Press).
O'Connor, M. (1994), 'On the Misadventures of Capitalist Nature', in M. O'Connor (ed.), *Is Capitalism Sustainable? Political Economy and the Politics of Ecology* (New York and London: Guilford), 125–51.
OECD (1987), *Pricing of Water Services* (Paris: Organization for Economic Co-operation and Development).
OECD (1999), *The Price of Water: Trends in OECD Countries* (Paris: Organization for Economic Co-operation and Development).
Offe, C. (1984), *Contradictions of the Welfare State* (London: Hutchinson).
—— (1985), *Disorganized Capitalism: Contemporary Transformations of Work and Politics* (Cambridge: Polity).
Offer, A. (2002), 'Why has the Public Sector Grown so Large in Market Societies? The Political Economy of Prudence in the UK, *c*.1870–2000', Discussion Papers in Social and Economic History, no. 44 (March), Oxford University.
OFT (1999), *The Competition Act 1998: Market Definition, OFT 403* (London: Office of Fair Trading).
Oftel (1985), *Annual Report: 5 August to 31 December 1984* (London: Office of Telecommunications/HMSO).
Ofwat (1990), *Paying for Water: A Time for Decisions*, Consultation paper on future charging policy for water and sewerage services (Birmingham: Office of Water Services).
—— (1991*a*), *Profits and Dividends: Letter to Managing Directors of Water and Sewerage Companies* (Birmingham: Office of Water Services).
—— (1991*b*), *Debt and Disconnection: Letter to Managing Directors of Water and Sewerage Companies* (Birmingham: Office of Water Services).
—— (1992*a*), *Paying for Water: The Way Ahead* (Birmingham: Office of Water Services).
—— (1992*b*), *Increasing Competition in the Water Industry* (Birmingham: Office of Water Services).
—— (1992*c*), *Guidelines on Optional Metering* (Birmingham: Office of Water Services).
—— (1993*a*), *Paying for Growth* (Birmingham: Office of Water Services).
—— (1993*b*), *Paying for Quality* (Birmingham: Office of Water Services).
—— (1995*a*), *1994–95 Report on the Financial Performance and Capital Investment of the Water Companies in England and Wales* (Birmingham: Office of Water Services).
—— (1995*b*), *Compensation for Customers:* Consultation paper on the issues raised by the1995 Drought (Birmingham: Office of Water Services).
—— (1996*a*), *Leakage of Water in England and Wales* (Birmingham: Office of Water Services).
—— (1996*b*), *Diversification by Water Companies* (Birmingham: Office of Water Services).
—— (1996*c*), *The Changing Structure of the Water and Sewerage Industry in England and Wales* (Birmingham: Office of Water Services).
—— (1996*d*), *Water Charges and Company Profits* (Birmingham: Office of Water Services).

Bibliography

—— (1996*e*), *Transfer Pricing in the Water Industry: A Review of the Activities of the Water and Sewerage Companies* (Birmingham: Office of Water Services).

—— (1996*f*), *Budget Payment Units for Water and Sewerage Charges* (Birmingham: Office of Water Services).

—— (1996*g*), *The Regulation of Common Carriage Agreements in England and Wales*, Consultation Paper (Birmingham: Office of Water Services).

—— (1996*h*), *Report on Conclusions from Ofwat's Enquiry into the Performance of Yorkshire Water Services Ltd* (Birmingham: Office of Water Services).

—— (1996*i*), *Customer Benefits, Investments and Dividends: Letter to Managing Directors of all Water and Sewerage Companies and Water Only Companies* (Birmingham: Office of Water Services).

—— (1996*j*), *Compensation for Customers: Review of the Guaranteed Standards Scheme* (Birmingham: Office of Water Services).

—— (1997*a*), *1996–97 Report on Leakage and Water Efficiency* (Birmingham: Office of Water Services).

—— (1997*b*), *1996–7 Report on the Financial Performance and Capital Investment of the Water Companies in England and Wales* (Birmingham: Office of Water Services).

—— (1997*c*), *Water Pricing: The Importance of Long-Run Marginal Costs* (Birmingham: Office of Water Services).

—— (1997*d*), *Supply/Demand Balance Submission: Reporting Requirements* (Birmingham: Office of Water Services).

—— (1997*e*), *1999 Periodic Review: A Letter to the Managing Directors of all Water and Sewerage and Water Only Companies* (Birmingham: Office of Water Services).

—— (1997*f*), *Annual Mandatory Leakage Targets: A Letter to the Managing Directors of all Water and Sewerage and Water Only Companies* (Birmingham: Office of Water Services).

—— (1997*g*), *The Proposed Framework and Approach to the 1999 Periodic Review*, Consultation paper (Birmingham: Office of Water Services).

—— (1998*a*), *1997–98 Report on Tariff Structure and Charges* (Birmingham: Office of Water Services).

—— (1998*b*), *Setting the Quality Framework: An Open Letter to the Secretary of State for the Environment, Transport and the Regions and the Secretary of State for Wales* (Birmingham: Office of Water Services).

—— (1998*c*), *Assessing the Scope for Future Improvements in Water Company Efficiency*, Technical paper (Birmingham: Office of Water Services).

—— (1998*d*), *Prospects for Prices: A Consultation Paper on Strategic Issues Affecting Future Water Bills* (Birmingham: Office of Water Services).

—— (1999*a*), *1999–2000 Report on Tariff Structure and Charges* (Birmingham: Office of Water Services).

—— (1999*b*), *Approval of Companies' Charges Schemes in 2000–2001*, Letter to Managing Directors (Birmingham: Office of Water Services).

—— (1999*c*), *Final Determinations: Future Water and Sewerage charges (2000–05)* (Birmingham: Office of Water Services).

—— (2000*a*), *2000–2001 Report on Tariff Structure and Charges* (Birmingham: Office of Water Services).

—— (2000*b*), *The Proposed Restructuring of the Kelda Group: A Consultation Paper by the Director General of Water Services* (Birmingham: Office of Water Services).

Ofwat (2000c), *New Ownership Structures in the Water Industry* (Birmingham: Office of Water Services).

—— (2000d), *The Proposed Restructuring of the Kelda Group: A Preliminary Assessment by the Director General of Water Services* (Birmingham: Office of Water Services).

—— (2000e), 'Kelda's Proposals to Restructure Yorkshire Water are not Acceptable, Says Sir Ian Byatt', Press release (Birmingham: Office of Water Services), 25 July.

—— (2001a), *Competition in the Provision of New Water Mains and Service Pipes*, Consultation paper (Birmingham: Office of Water Services).

—— (2001b), *The Proposed Acquisition of Mid-Kent Holdings plc by Swan Capital Investments*: A Consultation paper (Birmingham: Office of Water Services).

—— (2001c), *The Proposed Acquisition of Dŵr Cymru Cyfyngedig by Glas Cymru Cyfyngedig*, Position paper (Birmingham: Office of Water Services).

—— (2002a), *The Water Framework Directive: Implications for Capital Investment*, Presentation by Philip Fletcher, Director General of Water Services to Water UK, Foresight Series Water Framework Directive Conference, 2 May (London: Water UK).

—— (2002b), *Market Competition in the Water and Sewerage Industry*, Information note no. 10 (Birmingham: Office of Water Services).

—— (2002c), *Financial Performance and Expenditure of the Water Companies in England and Wales* (Birmingham: Office of Water Services).

Ogden, S. (1991), 'The Trade Union Campaign Against Privatisation: Industrial Impotence or Political Maturity?' *Industrial Relations Journal*, 11/1: 20–35.

—— (1997), 'Accounting for Organizational Performance: The Construction of the Customer in the Privatized Water Industry', *Accounting, Organizations and Society* 22 (6), 529–556.

—— and Glaister, K. (1996), 'The Cautious Monopolists: Strategies of Britain's Privatized Water Companies', *Long Range Planning*, 29/5: 663–74.

Okun, D. A. (1977), *Regionalisation of Water Management: A Revolution in England and Wales* (London: Applied Science Publishers).

Olson, M. (1965), *The Logic of Collective Action: Public Goods and the Theory of Groups* (Cambridge, Mass.: Harvard University Press).

—— (1982), *The Rise and Decline of Nations: Economic Growth, Stagflation and Social Rigidities* (New Haven: Yale University Press).

O'Neill, J. (1997), 'Value Pluralism, Incommensurability and Institutions', in J. Foster (ed.), *Valuing Nature? Ethics, Economics and the Environment* (London: Routledge), 75–88.

ONS (1998) *National Accounts Concepts, Sources and Method* (London: HMSO).

Orange, M. (2003), 'Suez renonce à sa stratégie d'expansion pour se recentrer sur ses métiers les plus rentables', *Le Monde*, Jan. p. 19.

Osborne, D. and Gaebler, T. (1993), *Reinventing Government: How the Entrepreneurial Spirit is Transforming the Public Sector* (New York: Plume Books).

Ostrom, E. (1990), *Governing the Commons: The Evolution of Institutions for Collective Action* (Cambridge: Cambridge University Press).

OXERA (1994a), *Guide to the Economic Regulation of the Water Industry* (Oxford: Oxford Economic Research Associates Ltd).

—— (1994b), *Acquisitions and Diversification: The Record of the Privatised Utilities* (Oxford: Oxford Economic Research Associates Ltd).

—— (1995), *Acquisitions and Diversification: The Record of the Privatised Utilities* (Oxford: Oxford Economic Research Associates Ltd).

—— (1996), *Acquisitions and Diversification: The Record of the Privatised Utilities* (Oxford: Oxford Economic Research Associates Ltd).

—— (1997), *Water Resources Demand* (Oxford: Oxford Economic Research Associates Ltd).

Page, B. (2002), 'Has Widening Participation in Decision-making Influenced Water Policy-making in the UK?', Working Papers in Economic Geography, WP 02-08, School of Geography and the Environment, Oxford University.

—— and Swyngedouw, E. (2002), 'Results of the Case Study on the Water Supply in the Region of London, UK', in *Achieving Sustainable and Innovative Policies Through Participatory Governance in a Multi-Level Context*, Final Report (Water), Research project funded by the European Community under the 5th Framework Programme (Improving Human Potential and Socio-economic Knowledge base), contract no. HPSE-CT-1949-00028 (Oxford: School of Geography and the Environment, Oxford University).

Panayotou, T. (1993), *Green Markets: The Economics of Sustainable Development* (San Francisco: Institute for Contemporary Studies Press).

Parker, D. (2000), 'Reforming Competition Law in the UK: The Competition Act 1998', Occasional Paper 14, Centre for the Study of Regulated Industries, School of Management, University of Bath.

—— and Penning-Rowsell, C. (1980), *Water Planning in Britain* (London: Allen & Unwin).

Patterson, A. and Pinch, P. L. (1995), '"Hollowing-out" the Local State: Compulsory Competitive Tendering and the Restructuring of British Public Services', *Environment and Planning A*, 27/9: 1437–61.

Pearce, D. (1976), *Environmental Economics* (London: Longman).

Pearce, F. (1982), *Watershed: The Water Crisis in Britain* (London: Junction Books).

Peck, J. and Tickell, A. (1994), 'Searching for a New Institutional Fix: the *After*-Fordist Crisis and Global–Local Disorder', in A. Amin (ed.), *Post-Fordism: A Reader* (Oxford: Blackwell), 280–315.

Peet, R. and M. Watts (1993), 'Introduction: Development Theory and Environment in an Age of Market Triumphalism', *Economic Geography*, 69: 227–53.

Penning-Rowsell, E. C. and Parker, D. J. (1983), 'The Changing Economic and Political Character of Water Planning in Britain', *Progress in Resource Management and Environmental Planning*, 4: 169–99.

Perera, M., Farley, A. *et al.* (1985), 'The London Water Ring Main: A New Approach to London's Water Distribution Strategy', *Aqua*, 6: 304–10.

Pickart, M. (2001), 'Fifty Years after Samuelson's "The Pure Theory of Public Expenditure"', paper presented at the 52nd Atlantic Economic Conference', http://www.iaes.org/conferences/future/philadelphia_52/prelim_program/h00-1/pickhardt.pdf. Accessed 30 Oct.

Pierre, J. (2000) (ed.), *Debating Governance* (Oxford: Oxford University Press).

Pierson, C. (1998), *Beyond the Welfare State: The New Political Economy of Welfare* (University Park, Penn.: Pennsylvania University Press).

Pigou, A. (1920), *The Economics of Welfare* (London: Macmillan).

Pirie, M. (1988), *Privatisation: Theory, Practice and Choice* (Aldershot: Wildwood House).

Porter, A. (1978), 'Equalization Explained', *Water: The Journal of the National Water Council* (Mar.), 3–4.
Power, M. (1994), *The Audit Explosion* (London: Demos).
Prusmann, D. F. (1968), 'The Finance of Public Utility Companies in the Water Supply Industry: Some Aspects', *Public Administration*, 46: 63–79.
Pugh, N. J. (1939), 'Municipal Trading: Water Supply', *Public Administration*, 17: 276–92.
Rajah, N. and Smith, S. (1993), 'Distributional Aspects of Household Water Charges', *Fiscal Studies*, 14/3: 86–108.
Rathbone P (2002), 'Don't Fence Me In', *Utility Week*, 18 January, 18–19.
Redwood, J. and Hatch J. (1982), *Controlling Public Industries* (Oxford: Basil Blackwell).
Rees, J. (1989), *Water Privatisation*. Research Papers, Department of Geography. LSE, London.
—— (1992), 'Markets: The Panacea for Environmental Regulation?', *Geoforum*, 23/3: 383–94.
—— (1998), 'Regulation and Private Participation in the Water and Sanitation Sector', *Natural Resources Forum*, 22/2, 95–105.
—— and Williams, S. (1993), *Water for Life: Strategies for Sustainable Water Management* (London: Council for the Protection of Rural England).
Reid, D. (1991), *Paris Sewers and Sewermen: Realities and Representations* (Cambridge, Mass.: Harvard University Press).
Research by Design (1996), *The Drought of 1995: The Customers' Perspective in Bradford*, Research study conducted on behalf of the Office of Water Services (Solihull: Research by Design).
Richardson, J. (1994), 'EU Water Policy: Uncertain Agendas, Shifting Networks and Complex Coalitions', *Environmental Politics*, 3/4: 139–67.
—— Maloney, W. and Rudig, W. (1992), 'The Dynamics of Policy Change: Lobbying and Water Privatization', *Public Administration*, 70: 157–75.
Risbridger, C. A. (1963), 'Compensation Water, Reuse of Water and Waste Prevention', in *Conservation of Water Resources* (London: Institution of Civil Engineers), 97–106.
Roberts, R. (1995), 'Taking Nature-Culture Hybrids Seriously in Agricultural Geography', *Environment and Planning A*, 27, 673–82.
—— and Emel, J. (1992), 'Uneven Development and the Tragedy of the Commons: Competing Images for Nature-Society Analysis', *Economic Geography*, 68/3: 249–71.
Robinson, C. (1997), 'Introducing Competition into Water', in M. E. Beesley (ed.), *Regulating Utilities: Broadening the Debate* (London: Institute for Economic Affairs in association with the London Business School).
Robson, W. A. (1935), 'The Public Utility Services', in H. Laski, I. Jennings, and W. A. Robson (eds.), *A Century of Municipal Progress, 1835–1935* (London: Allen & Unwin), 299–331.
Rofe, B. H. and Hampson, E. (1995a), *Yorkshire Water Drought Orders 1995: Independent Assessment Report* (London: Department of the Environment).
—— —— (1995b), *Yorkshire Water Drought Orders 1995: Second Assessment Report* (London: Department of the Environment).
Rogers, P. (1993), *America's Water: Federal Roles and Responsibilities* (Cambridge, Mass.: MIT Press).

Rose, C. (1991), *The Dirty Man of Europe: The Great British Pollution Scandal* (London: Simon and Schuster).

Ross, A. (1994), *The Chicago Gangster Theory of Life* (London and New York: Verso).

Rowson, I. (2000), 'The Design of Competition in Water', Occasional paper 15, Centre for the Study of Regulated Industries, University of Bath School of Management.

RPA (1998), *The Environmental Costs and Benefits of Water Resources: A Preliminary Methodology* (Bristol: Environment Agency).

Russac, D., Rushton, K. *et al.* (1991) 'Insights into Domestic demand from a Metering Trial', *Journal, Institution of Water and Environmental Management*, 5/3: 342–51.

Saal, D. and Parker, D. (2001), 'Productivity and Price Performance in the Privatized Water and Sewerage Companies of England and Wales', *Journal of Regulatory Economics*, 20/1: 61–90.

Sagoff, M. (1988), *The Economy of the Earth: Philosophy, Law and the Environment* (Cambridge: Cambridge University Press).

Samson, C. (1994), The Three Faces of Privatisation, *Sociology*, 2/1: 79–97.

Samuelson P. (1954), 'The Pure Ttheory of Public Expenditure', *Review of Economics and Statistics*, 36: 387–9

Saunders, P. and Harris, C. (1990), 'Privatization and the Consumer', *Sociology*, 24: 57–75.

—— —— (1994), *Privatisation and Popular Capitalism* (Buckingham and Philadelphia: Open University Press).

Sawkins, J. W. (1995), 'Yardstick Competition in the English and Welsh Water Industry: Fiction or Reality?', *Utilities Policy*, 5/1: 27–36.

SCF (1996), *Water Tight: The Impact of Metering on Low Income Families* (London: Save the Children Fund).

Schoenberger, E. (1991*a*), 'The Corporate Interview as a Research Method in Economic Geography', *Professional Geographer*, 43/2: 180–9.

—— (1991*b*), 'Self-Criticism and Self-Awareness in Research: A Reply to Linda McDowell', *The Professional Geographer*, 44/2: 215–18.

Schofield, R. and Shaoul, J. (1996), 'Regulating the Water Industry: Swimming Against the Tide or Going Through the Motions?', Public Interest Report, University of Manchester (Aug.).

—— —— (1997), 'Regulating the Water Industry: Swimming Against the Tide or Going through the Motions?' *Ecologist*, 27/1: 6–13.

Schramm, E. and Kluge T. (1994), 'The German Water Crisis: A Socio-Ecological View', *Capitalism, Nature, Socialism*, 5/1: 99–113.

Scott, S. (1995), 'LRMC and Charging the Polluter: The Case of Industrial Waste Water in Ireland', *Utilities Policy*, 5/2: 147–64.

Scottish and Westminster Communications (1992), *The Implications of Privatisation of Water and Sewerage Services in England and Wales* (Edinburgh: Convention of Scottish Local Authorities).

Sen, A. (1999), *Commodities and Capabilities* (New Delhi: Oxford University Press).

Shaoul, J. (1997), 'A Critical Financial Analysis of the Performance of Privatized Industry: The Case of the UK Water Industry', *Critical Perspectives on Accounting*, 8: 479–505.

—— (2000*a*) 'Water Mutualisation: The Financial Reality of the New Form of Ownership—Response to Kelda's Proposal to Sell the Regulated Water Business to a Mutual Owned by Consumers', MS, School of Accounting and Finance, Manchester University.

Shaoul, J. (2000b), 'Tapping into Mutuals', *Public Finance* (8–14 September), 16–18.

Sheail, J. (1982), 'Underground Water Abstraction: Indirect Effects of Urbanization on the Countryside', *Journal of Historical Geography*, 8/4: 395–408.

—— (1983), 'Planning, Water Supplies and Ministerial Power in Inter-War Britain', *Public Administration*, 61: 386–95.

—— (1986), 'Government and the Perception of Reservoir Development in Britain: An Historical Perspective', *Planning Perspectives*, 1: 45–60.

—— (1995), 'New Towns, Sewerage and the Allocation of Financial Responsibility: The Post-war UK Experience', *Town Planning Review*, 66/4: 371–87.

Shiva, V. (2002), *Water Wars: Privatization, Pollution and Profit* (Boston: South End Press).

Sidgwick, H. (1874), *The Methods of Ethics* (London: Macmillan).

Simon, O. (1986), 'Investing in the Infrastructure', *National Westminster Bank Quarterly Review* (May), 2–17.

Sleeman, J. F. (1953), *British Public Utilities* (London: Pitman).

Smith, K. (1972), *Water in Britain: A Study in Applied Hydrology and Resource Geography* (London: Macmillan).

Smith, N. (1984), *Uneven Development* (Oxford: Blackwell).

Snow, J. (1936, orig. 1855), *On the Mode of Communication of Cholera* (New York: Commonwealth Fund).

Social Enterprise Institute (2001), 'Scottish Water: The Case for a People's Company with Mutual Ownership and Management', Heriot Watt University, Social Enterprise Institute and Scottish Co-operative and Mutual Forum. June. http://www.som.hw.ac.uk/socialenterprise/Reports.html. Accessed 16 Oct.

Solway, J. S. (1994), 'Drought as a "Revelatory Crisis": An Exploration of Shifting Entitlements and Hierarchies in the Kalahari, Botswana', *Development and Change* 25: 471–495.

Stiglitz, J. E. (2002), *Globalization and its Discontents* (London and New York: W. W. Norton).

Strang, V. (2001), 'Evaluating Water: Cultural Beliefs and Values about Water Quality, Use and Conservation' (London: Water UK).

Summerton, N. (2001), 'Changes to the Institutional Structure of Water in England and Wales?', *MIA News: Newsletter of the International Water Association Specialist* (April), 22–9.

Swyngedouw, E. (1989), 'The Heart of the Place: The Resurrection of Locality in the Age of Hyperspace', *Geografiska Annaler*, 71 (b): 31–42.

—— (1992), 'Territorial Organization and the Space/Technology Nexus', *TIBG*, 17/4: 417–33.

—— (1997), 'Power, Nature, and the City: The Conquest of Water and the Political Ecology of Urbanization in Guayaquil, Ecuador', 1880–1990, *Environment and Planning A*, 29: 311–32.

—— (1999), 'Modernity and Hybridity: Nature, *Regeneracionismo*, and the Production of the Spanish Waterscape, 1890–1930', *Annals of the Association of American Geographers*, 89/3: 443–65.

Synnott, M. F. (1985), *The Relationship between the Regional Water Authorities and Local Planning Authorities*, unpublished doctoral thesis, Department of Geography, University College/ London School of Economics, London.

Tabony, R. C. (1977a), 'Drought Classification and a Study of Drought at Kew', *Meterological Magazine*, 106: 129–45.
—— (1977b), *The Variability of Long-Duration Rainfall over Great Britain*, Scientific paper no. 37 (London: Meteorological Office).
Taylor, A. (1998), 'Water Groups Warned', *Financial Times*, 8 June, 3.
—— and Pickard, J. (2001), 'People's Company' in Water Deal', *Financial Times* online: January 31. Accessed 5 May.
Thackray, J. (1995), *New Bills for Old: The Dilemmas of Water and Sewerage Charges*, Discussion paper no. 12 (London: Centre for the Study of Regulated Industries).
Thames Water (1991), *Annual Report and Accounts* (Reading: Thames Water).
—— (1992), *Annual Report and Accounts* (Reading: Thames Water).
—— (1993), *Annual Report and Accounts* (Reading: Thames Water).
—— (1994), *Annual Report and Accounts* (Reading: Thames Water).
—— (1996), *Annual Report and Accounts* (Reading: Thames Water).
—— (1997), *Annual Report and Accounts* (Reading: Thames Water).
Thomasson, K. (1993), *Keynote Report: Water Utilities* (Hampton, Middx: Keynote Publications).
Tullock, G. (1976), *The Vote Motive: An Essay in the Economics of Politics, with Applications to the British Economy* (Princeton: Princeton University Press).
Turton, P. (1995), *Domestic Consumption Monitoring Survey* (Worthing: NRA Demand Management Centre).
Twort, A. C. (1963), *A Textbook of Water Supply* (London: Edward Arnold).
Uff, J. (1996), *Water Supply in Yorkshire: Report of the Independent Commission of Inquiry* (Leeds: Yorkshire Water Services Ltd.).
UKWIR/NRA (1995), *Demand Forecasting Methodology: Main Report* (London: UK Water Industry Research/National Rivers Authority).
Usher, A. D. (ed.) (1997), *Dams as Aid: A Political Anatomy of Nordic Development Thinking* (London: Routledge).
Utilities Journal (2000a), 'The Glas Cymru Proposals' (Dec.), 34–5.
—— (2000b), 'The Commission's Decisions' (Oct.), 32–3.
—— (2001a), 'Gearing up for Bond Issues' (April), 30–1.
—— (2001b), 'The Glas Decision' (Feb), 28–9.
—— (2001c), 'Taking the Long View' (June), 30–1.
—— (2001d), 'The Main Event?' (Sept.), 28–9.
Utility Week (1995), 'IT in Utilities', Special issue (19 Nov.), 13–15.
Vass, P. (2002) 'Competition and Restructuring in the Water Industry', *Journal of Network Industries*, 3: 77–98.
Vickers, J. (1997), 'Regulation, Competition and the Structure of Prices', *Oxford Review of Economic Policy*, 13/1: 15–26.
Vickers, J. and Yarrow, G. (1988), *Privatization: An Economic Analysis* (London: MIT Press).
von Tunzelmann, G. N. (1978), *Steam Power and British Industrialisation to 1860* (Oxford: Oxford University Press).
Walker, D. L. (1983), 'The Effect of European Community Directives on Water Authorities in England and Wales', *Aqua*, 4: 145–7.
Walsh, K. (1995), *Public Services and Market Mechanisms: Competition, Contracting and the New Public Management* (New York: St. Martin's Press).

Walzer, M. (1983), *Spheres of Justice: A Defence of Pluralism and Equity* (London: Basic Books).
Ward, C. (1997), *Reflected in Water* (London: Cassell).
Ward, N. (1996), 'Surfers, Sewage and the New Politics of Pollution', *Area*, 28/3: 331–8.
Water Briefing (1995), ' "Water Only" plcs Make Point of Drought Records in Interim Result Reports', 9–11.
—— (1996*a*), 'Wessex Water Leaves Industry Association', 7 Feb., 15.
—— (1996*b*), 'Fat Cats Join Lower Ranks in Latest Staff Shakeouts', 3 April, 8.
WaterWatch (1996), *Submission of Evidence to the House of Commons Environment Select Committee Inquiry into Water Conservation and Supply* (Rotherham: Waterwatch).
West, B. and Smith, P. (1996), 'Drought, Discourse and Durkheim: A Research Note', *Australian and New Zealand Journal of Sociology*, 32/1: 93–102.
WHICH (1996), *Don't Pour Money Down the Drain* (London: WHICH/Consumer's Association).
Whiteley, T. (1977), 'Paying for Water—by Rates or Charges?' *National Westminster Bank Quarterly Review* (Aug.), 31–40.
Winpenny, J. (1994), *Managing Water as an Economic Resource* (London: Routledge).
—— (1997), *Water Policy Issues*. DFID Water Resources occasional papers. (Loughborough and London: Water and Engineering Development Centre and Department for International Development).
Wolfe, J. (1991), 'State Power and Ideology: Mrs Thatcher's Privatization Programme', *Political Studies*, 39.
World Bank (1997), *The State in a Changing World: World Development Report, 1997* (Oxford: Oxford University Press).
World Bank/IFC (1991), *Office Memorandum: Privatization of the Water Sector in the UK and France* (Washington: World Bank/International Finance Corporation).
World Commission on Dams (2001), *Dams and Development: A New Framework for Decision-Making* (London: Earthscan).
WRB (1974), *Water Resources in England and Wales* (London: Water Resources Board).
WSA (1987), *Waterfacts '86* (London: Water Services Association).
—— (1995),*Waterfacts '94* (London: Water Services Association).
—— (1997), *Water: Meeting the Challenge: The Vision of the Water and Sewerage Companies of England and Wales* (London: Water Services Association).
—— (1998), *Waterfacts '97* (London: Water Services Association).
—— (2000), *Waterfacts 2000* (London: Water Services Association).
Wynne, B. (1994), 'Scientific Knowledge and the Global Environment', in M. Redclift and T. Benton (eds.), *Social Theory and the Global Environment* (London and New York: Routledge).
Xenos, N. (1989), *Scarcity and Modernity* (London and New York: Routledge).
Yorkshire Post (1995), 'Firms Hit by Water Shortages . . . should move', 29 Aug., 1.
Yorkshire Water, (1995), *Annual Report and Accounts* (Leeds: Yorkshire Water plc).
—— (1997), *Establishing the Economic Level of Leakage* (Leeds: Yorkshire Water Services Ltd.).
—— (2000*a*), 'Yorkshire Water's Future: What Does It Mean For The Customer?' Press release, 23 June (Leeds: Yorkshire Water plc).
—— (2000*b*), *Kelda Proposals for the Future of Yorkshire Water* (Leeds: Yorkshire Water plc).

—— (2000c), *Kelda Group Strategy Review: Conclusions* (Leeds: Yorkshire Water plc).
Yorkshire Wildlife Trust (1996), *Yorkshire Wildlife Trust Objection to the River Hull Drought Permit Application by Yorkshire Water Services Ltd.* (York: Yorkshire Wildlife Trust).
Young, I. M. (1990), *Justice and the Politics of Difference* (Princeton: Princeton University Press).
Zarnikau, J. (1994), 'Spot Market Pricing of Water Resources and Efficient Means of Rationing Water during Scarcity (Water Pricing)', *Resource and Energy Economics*, 16/3: 189–210.
Zimmerer, K. S. (1993), 'Soil Erosion and Social (Dis)courses in Cochabamba, Bolivia: Perceiving the Nature of Environmental Degradation', *Economic Geography*, 69: 312–28.
—— (1994), 'Human Geography and the New Ecology: The Prospect and Promise of Integration', *Annals of the Association of American Geographers*, 84/1: 108–25.

INDEX

Abstraction licence 75
Anglian Water 80, 158, 174–5
Asset management plans 112–14
Arms-length regulation, *see* regulation
Axford case 82–4
Averch-Johnson effect 148

Benton, T. 32
British Medical Association 131
British Water 162, 164
Byatt, Sir Ian 176
British model 3–9, 17, 68–72, 111–14, 147, 154, 175–8, 186–7
  see also price caps; price review; regulation

Canada 46, 162
capital expenditure 112–13, 176
  forecast requirements 8, 64
  post WW II 55–6, 62–3
climate change 87
commercialization 6–7, 13–14, 16, 25, 33, 39, 42, 66, 72–3, 74–98, 123, 155, 182, 189
competition 5, 51, 75–6, 79, 92–7, 138, 179
  common carriage 96–7
  inset appointments 95–6
Competition Act (1998) 149
Competition Commission 70, 93, 169
connections 14, 21, 43–4, 46–7, 50
  see also disconnections
Conservative Government/Party 3, 8, 62, 65, 67–8, 126–7, 134, 136, 150
consumers:
  debt 131–3
  identity as citizens/consumers 21–2, 26, 182, 187–9
  low-income 129–33, 136–9, 141
  and water pricing 123–44
  see also water poverty
Consumer's Association 80, 133
conservation 26–7, 31, 75, 90–2, 105
  See also demand management; water
Contingent valuation: *see* environmental valuation
Council for the Protection of Rural England 70
cross-subsidies 20, 21, 91–2, 94–5, 123–8, 135–6, 137–9

dams 19, 23, 45
demand:
  forecasts 111–13
  levels 23, 65
  management 5, 27, 87, 90–2, 113–14
  see also conservation; water
disconnections 132–3, 136
discourse and regulation 38–9, 120–1
Department for International Development 164
Department of the Environment, Food, and Rural Affairs 62, 68, 70, 82–3, 96, 135, 137, 156, 164–5
Department of the Environment, Transport, and Regions, *see* Department of the Environment, Food, and Rural Affairs
Department of the Environment, *see* Department of the Environment, Food, and Rural Affairs
Department for Trade and Industry 163–5
Director General of Water Services, *see* Office of Water Services
dividends 105, 148, 150–1, 152–3, 161–2
  see also profits
diversification 16, 155–62
Drinking Water Inspectorate 69, 70
drought:
  orders 105–10
  planning 119
  Yorkshire region 15, 101–22
  see also Yorkshire Water
Dwr Cymru 142
  mutualization proposal 10, 145, 171–4, 176

ecological fix 35, 39
ecological modernization 27, 39
efficiency 3, 5–6, 12–15, 26–7, 31, 48, 51, 60, 75, 76–9, 89–92, 105, 112–13, 117, 119–21, 123, 127–8, 139–41, 145, 147–8, 151, 153–5, 171, 173–4, 176, 179–81, 183, 185, 188, 190
equalization 16, 123–8, 137
  see also consumers; prices
equity:
  definitions 27, 124, 190–2
  and water pricing 16, 26, 27, 123–44, 179
employment levels 5, 66, 117–18, 155
Environment Act (1995) 86
Environment Agency 68, 70, 81–7, 103, 106, 111, 116, 119, 129, 135, 149, 151

environmental management 4–5, 12–16, 18–41, 45, 50, 63–4, 70–1, 81–6, 88, 97, 179, 182–3, 192–5
  see also water; water quality
environmental valuation 12, 13, 74–6, 181–6
European Commission, see European Union
European Union 64, 69
externalities 22, 29, 31–6, 39, 75, 76

Fletcher, Steven 176
France 46, 129
French water companies 21
Friends of the Earth (UK) 70
Fuel poverty 141

Gas and Water Facilities Act (1870) 52
Glas Cymru, see Dwr Cymru
Guaranteed Standards Scheme 115, 150

Harvey, D. 28, 32, 33–4
House of Commons Environment Committee 169

industrialization 42, 43–4
Integrated River Basin Management 60
internationalization 16, 156, 162–7, 185–7
inset appointments, see competition

Kelda, see Yorkshire Water

labour, see employment levels
Labour Government/Party 58, 62, 119, 125–6, 133, 135–6, 150, 187, 191–2
leakage 83, 88, 90–1, 101, 104–6, 108, 116–17, 119, 167, 180, 184
long run marginal cost 89–91, 138, 190
light touch regulation, see regulation
Littlechild, S. 135, 180
Lyonnaise des Eaux 178

market environmentalism 12, 14, 18–19, 23–8, 39–40, 75, 148, 193–5
market failure 5, 10, 22–3, 24–5
materiality, theory of 31–3
metering 5, 13–14, 21, 27, 56, 74–6, 79–81, 86–9, 90–1, 94–5, 98, 123–4, 128–9, 131–3, 134, 136–8, 150, 190
  budget payment units 133, 139
  optional policy 79–81
  rates of 5, 146
  and vulnerable consumers 129, 136
mergers 158–61
Monopolies and Mergers Commission, see Competition Commission

municipal ownership/management of water supply 20–1, 48–9, 51–8
mutualization 187–9, 169–71
  see also Dwr Cymru (Welsh Water)
  see also Kelda (Yorkshire Water)

National Consumer Council 133
National Rivers Authority 68, 87
  see also Environment Agency
nationalization 5, 58–67
natural monopoly 22
network utilities, see public utility services
non-governmental organizations 70
  environmental 26, 70
  consumer advocacy 70
North West Water 119
  see also United Utilities
Northumbrian Water 178

Offer, A. 45–6
Office of Fair Trading 149
Office of Water Services 86, 88, 104, 111, 116, 119, 127, 131, 133, 135, 147–8, 151, 161, 169–71, 173–4
  Customer Councils/Service Committees 70–1, 173
  duties of regulator 69, 78–9
  operating expenditure 112–14, 155
outsourcing 155
Overseas Development Aid 163
  see also Department for International Development

Pennon 174
periodic review, see price review
political ecology 40–1, 192–5
price caps 69, 112–14, 135
  evolution of, post-privatization 130–1, 134–6, 147–9, 151–4
  formula for determination of 78
  levels of (1999) 152
  regulation 78–9, 180–2
  see also British model; Ofwat; Littlechild; regulation
pricing 123–44, 183
  domestic consumers 5, 16, 142–4
  policy 16, 56, 77
  and Retail Price Index 77
price reviews 113–15, 116, 119, 135, 140, 142, 147, 149–53, 167, 176–7, 180–1, 183
private water companies, Victorian era 42–3, 46–51, 53
privatization:
  cost to government 68

Index

environmental dimension 12, 63–7
historical context 18, 179
theories of 7, 10–11, 13, 36–9, 39–40
stock market flotation of water industry 7–8, 67–8
profits 7, 42, 56–7, 68, 78, 94, 97, 102, 106, 112–14, 121, 135, 140, 147–55, 165, 175, 177, 180–1, 190, 193
  see also dividends; rates of return; share prices
Public Sector Borrowing Requirement 64–5, 67, 77
public good 22, 84–5
Public Health Acts (1848, 1875, 1878) 52, 54
public utility services 8, 10, 51–5, 56, 124, 139–40, 193
  see also externalities; industrialization; natural monopoly

rates of return 148, 151–2
  see also profits
regional water authorities 58–60, 62–7, 72, 77, 124–7, 156, 163
regionalization 58–67, 125
regulation:
  arms length 8, 69
  regulatory framework 68–71
  theory of 36–9
  see also Ofwat
re-regulation 3–4, 11–12, 15–16, 36–9, 118–21, 146–54, 179–82
Restructuring:
  and water supply industry post-privatization 16, 116–18, 140, 145–78
  of water supply sector, mid-twentieth century 54–8
  see also diversification; internationalization; mutualisation; nationalization; regionalization; vertical deintegration

Save the Children Fund 131
scarcity 26, 27, 28–9, 30–1, 34–5, 81–8, 121–2
  areas of official designation 88
Severn-Trent 129, 158, 174–5
share:
  prices 8, 153, 168
  ownership 8
  see also dividends; profits; rates of return
South Africa 22
South East Water 129
South West Water 128, 139, 142
Southern Water 93, 174
Spain 21

state failure 6, 15, 24–6
state hydraulic water management 12, 18–24, 28, 39–40, 76
Sutton & East Surrey Water 129, 168

tariffs, *see* pricing
Thames Water 82–3, 128–9, 165–7

uneven development 34
United States 21, 162
United Utilities 174–5
  see also North West Water
universal service obligation 5, 20, 51, 77
urbanization 42, 43–4

vertical deintegration 16, 156, 167–74, 177
Vivendi 93

Walzer, M. 21
water:
  biophysical characteristics and networked water supply 32–3, 60, 89, 94
  management, France 46
  management, Spain 20, 21, 46
  management, US 20, 46
  politics 183–5, 134–6
  and 'right to water' 21–2
  supply networks 42–3
  see also water quality
Water Act (1945) 55
Water Act (1973) 125, 127
Water Act (1983) 60, 77, 163, 166, 189
Water Authorities Association, *see* Water UK
Water Companies Association, *see* Water UK
Water Services Association, *see* Water UK
Water Framework Directive 69, 177
  see also European Union
Water Industry Act (1989, 1991, 1999) 88, 95, 132, 136, 139
Water Resources Act (1963) 57
water poverty 5, 44, 51, 54–5, 59–60, 64–5, 131–3, 141, 149, 184
Water quality 5, 12, 28, 48, 51–2, 56–7, 59–60, 62–5, 68–9, 72–3, 77, 79, 82, 86–7, 97, 101, 106, 114, 116, 118, 121–2, 128–9, 131–2, 134, 177, 180, 183–4, 194
  see also water
water resources, *see* water
Water Summit 119, 135, 150
Water supply, *see* water
Wateraid 164

Watervoice, *see* Ofwat
Waterwatch 9
Waterworks Clauses Act (1847) 50
Welsh Water, *see* Dwr Cymru
Wessex Water 174
Western Power 145

Windfall Tax 135, 150
World Bank 112, 164

Yorkshire Water 158
   mutualization proposal 9, 169–72
   and drought 15, 80, 101–22